Big Data in Organizations and the Role of Human Resource Management

PERSONALMANAGEMENT UND ORGANISATION

Herausgegeben von Volker Stein

Band 5

Zur Qualitätssicherung und Peer Review der vorliegenden Publikation

Die Qualität der in dieser Reihe erscheinenden Arbeiten wird vor der Publikation durch den Herausgeber der Reihe geprüft.

Notes on the quality assurance and peer review of this publication

Prior to publication, the quality of the work published in this series is reviewed by the editor of the series.

Tobias M. Scholz

Big Data in Organizations and the Role of Human Resource Management

A Complex Systems Theory-Based Conceptualization

PL ACADEMIC RESEARCH

Bibliographic Information published by the Deutsche Nationalbibliothek
The Deutsche Nationalbibliothek lists this publication in the Deutsche Nationalbibliografie; detailed bibliographic data is available in the internet at http://dnb.d-nb.de.

Zugl.: Siegen, Univ., Diss., 2016

Library of Congress Cataloging-in-Publication Data
Names: Scholz, Tobias, author.
Title: Big data in organizations and the role of human resource management : a complex systems theory-based conceptualization / Tobias M. Scholz.
Description: New York : Peter Lang, [2017] | Series: Personalmanagement und Organisation ; Vol. 5 | Includes bibliographical references.
Identifiers: LCCN 2016059623
Subjects: LCSH: Personnel management–Research. | Big data. | System theory.
Classification: LCC HF5549.A27 S36 2017 | DDC 658.4/03801–dc23 LC record available at https://lccn.loc.gov/2016059623

D 467
ISSN 1868-940X
ISBN 978-3-631-71890-2 (Print)
E-ISBN 978-3-631-71903-9 (E-PDF)
E-ISBN 978-3-631-71904-6 (EPUB)
E-ISBN 978-3-631-71905-3 (MOBI)
DOI 10.3726/b10907

PL Academic Research is an Imprint of Peter Lang GmbH.

Peter Lang – Frankfurt am Main · Bern · Bruxelles · New York · Oxford · Warszawa · Wien

This publication has been peer reviewed.

www.peterlang.com

Preface

In an environment where digitization permeates both society and economy at an ever-increasing pace, big data rank among the most fascinating challenges for all types of organizations. And their influence is not limited to those organizations concerned with matters of political administration such as national intelligence. All types of organizations, and especially those seeking to make a profit, i.e. companies, resort to big data. They now sense that the use of big data simultaneously entails fascinating opportunities and great risk.

Companies are immediately affected by the sheer momentum of the challenge that is big data, thus facing a series of profound questions: Do we even want to deal with big data? If so, what exactly do we want to do? What is possible, what is legal, what is reasonable, what is effective, and what can we legitimize? Those aspects refer to strategic decisions and, consequentially, to the more detailed questions regarding the actual execution of big data projects.

Tobias M. Scholz tackles exactly this in his dissertation. Even just consecutively reading through the array of practical as well as theoretical deficits he explicitly elaborates, reveals the overall chain of arguments: research on big data rarely dedicates itself to the human perspective – besides being a technological phenomenon, big data is also a social one – research rarely contextualizes big data towards particular corporations – big data challenge the role of the HR department – neither organizational theory nor theory on HR management adequately discuss the subjectivity of big data – research widely ignores the catalyzing effect of big data on complexity – the effects of big data on employees and the company are unclear – big data are insufficiently categorized theoretically – research on big data still lags behind in terms of practical application. Especially when putting into consideration those undeniable deficits, the subject of big data in organizations and the role of human resource management appear both pressing and highly economically relevant; above all doing so by means of a complex systems theory-based conceptualization.

For his dissertation, *Tobias M. Scholz* thus choses a topic that bares the potential for substantial innovation in both theory and practical application. In his work, he clearly utilizes said potential by initiating important developments in three distinct ways:

First of all, he provides a novel, concise, scientifically exact, and up-to-date outline, thus answering the question: "What are big data?" His very differentiated conceptualization goes beyond picking up numerous definitions and the evolutions thereof or differentiating said definitions from related concepts. He interconnects diverse developments of data-driven digitization and adjusts them to one another. In reference to a systematization introduced by *Boyd* and *Crawford*, he does away with unrealistic expectations regarding big data, thus bringing the concept back down to earth. He conducts a broad philosophical categorization of big data that includes critical observation. All things considered, he successfully illustrates the limitations of big data in organizations, while providing crucial hints as to how big data can be utilized sensibly. Especially the critical evaluation of those terms common to the big data discourse that are oftentimes used in a diffuse fashion, as well as of the only roughly implied paradigmatic progress, forms the base of his constructivist composition of alternative explanations and design suggestions.

Secondly, he stays true to his aim of specifying the implications of big data for organizations in general, and for the role of the HR department in particular. Not only does he successively walk the reader through his coherent mental framework; he integrates concepts derived from diverse strands of theory, among which being the ideas of complex systems theory, population ecology theory, and sociohistorical technology assessment, with their practical application. Particularly convincing is his differentiation between reactive, reactive-anticipating, and proactive roles of the HR department. In this context, he competently tackles future tasks that have arisen following the emergence of big data. Among those tasks are "big data risk governance" or "big data immersion", both featuring a strong link to HR economical practice, like that of "big data literacy" to HR development. En passant, he manages to develop a sustainable future role for the HR department, a corporate function that, as a result of the digitization and the pressing need for legitimization, finds itself at risk of being marginalized in corporate practice.

Thirdly, *Tobias M. Scholz* goes beyond elaborating a merely theory-based conceptualization on how to handle big data in organizations and the HR management. He also illustrates, in a differentiated manner, their implementability. He does so, firstly, with regards to practical application, by suggesting to fundamentally transform the HR department, while at the same time anticipating the emotional discussion and resistance this would entail, and sounding a word of caution when professionally handling this transformational challenge (of which he also provides a rough outline). He does so, secondly, with regards to research, by placing particular emphasis on social and ethical research challenges, stimulating further research on the transfer between theory and practice, and encouraging the HR research community to more intensely attend to novel paradigms such as gamification. He does so, thirdly, with regards to didactics in academia, by showing that both big data and the consequences of their application are fields of major didactic relevance.

More indirectly, *Tobias M. Scholz* takes a step in the theoretical discourse towards converging the logic of stabilization and that of dynamization. His major contribution is located on the conceptual metalevel. He bridges the gap between the necessities of constant organizational dynamization on the one hand, and the need for organizational balance on the other, thus postulating what he calls the "homeodynamic organization." His request to place the responsibility of creating such a coherence into the hands of the HR department lies grounded in the fact that the HR department is the only corporate function concerned with both employees as well as their data-related working conditions.

On the one hand, this dissertation about the interaction between data-related technology and human actors reveals to the reader that the implementation of big data is going to fundamentally change the corporate function of human resource management, as well as the way this transformation will occur. On the other hand, it illustrates the disposition of HR management itself to be more active, create more value, and drive the ethical implementation of big data in organizations. The fact that *Tobias M. Scholz* received this year's best dissertation award of the University of Siegen ("Förderpreis der Dirlmeier-Stiftung"), further attests to the excellence of his research.

Siegen, November 2016 *Univ.-Prof. Dr. Volker Stein*

Acknowledgement

What is the similarity between big data and the number 42 in the Hitchhiker's Guide to the Galaxy? Both give answers, but the questions are unknown.

Big data are complex and big data surround us, therefore, I am truly grateful to my doctoral adviser Volker Stein of allowing me to tackle such a research behemoth. Furthermore, giving me guidance and above all giving me the area of freedom to deal with this topic. I also want to thank Hanna Schramm-Klein and Arnd Wiedemann for being part of my thesis committee and for allowing me to present my thesis in such length and scope.

A special thank you goes to Florian Weuthen for his patience to read through my manuscript. The same goes to my mom, my father and my brother for their comments and feedback. A special dedication is to my cat Frodo; she could not see the end of this journey. On the way there were many colleagues who made the work a joy: Kevin Chaplin, Anna Feldhaus, Cornelia Fraune, Brigitte Grebe, Lena Kiersch, Martin F. Reichstein, Matthis S. Reichstein, Lina Ritter, Katrin Rödel, Katharina von Weschpfennig and Svenja Witzelmaier.

The book may be finished, but big data will become even more important in the upcoming years, so it will be interesting to see how we will transform big data and how big data will transform us.

Siegen, November 2016 *Tobias M. Scholz*

Table of Contents

List of Figures

List of Tables

1. Introduction

1.1 Statement of the Problem

01101000 01100101 01101100 01101100 01101111 00100000 01110111 01101111 01110010 01101100 01100100

In today's world we are surrounded by a vast amount of zeros and ones. Those numbers, called binaries and based on the Boolean logic, are the common way for computers to communicate with each other (Weyrich et al. 2014). Any digital information can be depicted in binary code. Machines and humans communicate in entirely different fashions since machines' communication is rooted within a computational logic (Kowalski 2011). Much like different languages, the communication between human and machine requires translation. The lines of numbers opening this chapter, for example, translated into writing spell "hello world", a familiar term for anyone studying a programming language. Both worlds (human world and machine world) seem to be separated with only few interconnections, but in recent years, the phenomenon of "big data" has led to an ongoing fusion of the two. Big data are here and big data are here to stay (Davenport 2014).

In an interplay of human and machine, both constantly generate new zeros and new ones. Be it me writing these words, a person talking on her/his phone, someone just walking the streets, be it a traffic light communicating with the control center, a smartphone logging into a Wi-Fi network, or a robot-arm in a factory. Every inhabitant of the modern world now inevitably leaves a digital trail, in addition, to her/his normal trail. The amount of data we generate on a daily basis is staggering, and increasing at an exponential rate. Everybody is participating in this digital world, contributing ever greater amounts of data. Most interestingly, nobody can hide from this digital trail. Even the decision of not participating at least generates the information of someone who does not wish to participate (Hartley & Chatterton 2001). In the future, the digital trail will become more precise, singular, and granular (Kucklick 2014) as there is an increase in the number of digital devices in the world as well as the frequency at which digital devices come to use (Economist 2014). Furthermore, the number of sensors that are capable of tracking people is constantly increasing and some sensors generate information about people as a by-product. These data only add to the already massive pile of existing data.

Although the information deluge (Hoenkamp 2012) is coupled with people leaving a more and more detailed digital footprint, the current development is predominantly driven by technology (Boyd & Crawford 2012). Big data pierce social life and society extensively. Society's solutions, however, are far from being precise or adequate. There is, in fact, a lack of social and ethical solutions (Barabási 2013). The need for a dispute of the social and ethical impact of big data is currently underestimated (e.g. Booch 2014), due to an inherent and precarious misconception. Big data are not the philosopher's stone (Crail 2015). Contrary to the opinion of some researchers (e.g. Anderson 2008), they are not ever going to reveal a certain and objective truth (Van Dijck 2014). Data are subjective, contextualized, heterogenic, and incomplete (Dalton & Thatcher 2014), while at the same time emitting an "aura of truth, objectivity, and accuracy" (Boyd & Crawford 2012: 664). Partly misled, humans overestimate the preciseness of big data due to the seeming objectiveness and become overconfident on the basis of data (Miller, C. C. 2015). However, shaped by such a narrative (Kosslyn 2015), big data narrow down the image of human behavior excessively and focus on standardized archetypes. Big data contribute to a "demystification of the world" (Weber 1919: 9). Interestingly, the technology behind big data, however, is currently being placed inside of a black box (Pasquale 2015), itself becoming something inscrutable (LaFrance 2015), something mystical (in analogy to Clarke 1977). At the very least, Drucker's statement (1967) that the computer is the moron, is no longer valid (Dewhurst & Willmott 2014).

There are reasons for outsourcing work and decisions to big data. In a complex world like the one we are living in, decisions need to be made in real time and under the pressure of a fluctuating and volatile environment which, therefore, makes constant change the new "stable" condition (Farjoun 2010). No human is capable of handling such massive complexity without the support of other humans and/or technological augmentations (Anderson & Rainie 2012). Big data are seemingly a technological enabler. Big data are a mixed blessing, supposedly capable of solving nearly any problem, but also the source of a staggering amount of new problems. Consequently, the mere use of big data will not suffice.

Deficit 1: Big data are not researched from a human perspective (Ekbia et al. 2015) and without a focus on the human factor (Zuboff 2014).

It is stated (Chen et al. 2012) that the usage of big data makes people's behavior more calculable and predictable. On the one hand, there is always the danger of employees believing that they are watched, much like a post-panopticon (Bauman 2000, Bakir 2015) or the electronic whip (West & Bowman 2014). This causes them to adapt their behavior. At first, big data may resemble Taylorism and could possibly lead to Taylorism 2.0 (deWinter et al. 2014), both with negative connotations, although at a second glance Taylorism has the benefit of being comprehensible. On the other hand, the algorithms behind big data are becoming increasingly unintelligible and potentially inaccurate (Kleinberg & Mullainathan 2015). From a technological perspective, we occupy a land of milk and honey where we can "gather data first, produce hypotheses later" (Servick 2015: 493). But, as Davis states, "Big data is pushing us to consider serious ethical issues including whether certain uses of big data violate fundamental civil, social, political, and legal rights" (2012: viii). This discourse is currently lagging behind the technological progress (Kitchin 2014a), despite the increasing significance of big data (Shaw 2014) and the "need of deeper critical engagement" (Crawford et al. 2014: 1664). Given the undeniable potential of big data to solve major problems, such discussion is of utmost importance.

Deficit 2: Big data are not purely technologically driven; they are a social phenomenon. However, the relation between big data and society is highly underresearched.

Big data are closely entangled with humans, as they only unfold their potential when utilized. Big data do not magically develop solutions and do not work independently from humans. It is, therefore, impossible to separate big data from human interaction. Big data may act as a black box. People may not understand the way big data work and may suspect they have a life of their own. Big data have a strong impact at the human level and will influence people drastically (Mayer-Schönberger & Cukier 2013). An interdisciplinary approach to this upcoming discussion is essential since the context of implications will vary. Situational environments determine the use of big data. Differences become obvious in the relationships between government and citizen (Kim et al. 2014), supplier and customer (Strong 2015), and employer and employee (Davenport 2014). Transferring big data strategy to another relationship without making contextual adjustments bears the danger of being inappropriate and even harmful. One field of human interaction is economic organization and, in particular, the usage of big data concerning employees. Employees as an integral part of an organization are neither enemies nor mere resources to be exploited, but rather an employers' partner with shared interests. This makes finding a potential competitive advantage for the company by adapting big data appropriately a delicate process. It might burden the trusting relationship between employer and employee. Marketing methods applying the shotgun principle are promising as they could lead to an increase in sales (Mayer-Schönberger & Cukier 2013), but using such methods with employees may disrupt the employer-employee relationship and harm commitment, performance, and retention.

Deficit 3: Big data are researched in a general way and not from a contextual viewpoint. In particular, the effects of big data in economic organizations are underresearched.

Within an organization, and especially within corporations, every use of big data will influence human relations (Harvard Business Review 2013). Even big data use in apparently nonadjacent fields will have an effect. The use of big data in research and development, for example, will lead to the creation of new products, and new products will impose different requirements of knowledge and skills onto employees. Big data are, therefore, bound to change work within organizations. One point of intersection of big data and humans to be considered is the human resource (HR) department. As a result of electronic human resource management (HRM), HRM have a long history of collecting and applying data. Using data in the analysis of employee relations is not a new turn, but the vastness of available data will represent a challenge to HRM. There is already a lot of information about the employees available to use (Kull 2016). It seems logical that not every individual member of an organization will handle big data but big data require steering by some entity within. The interests of both employers and employees will be incorporated into the use of big data. Consequently, big data as a technology will be driven by the IT department, however, as a social and human phenomenon will be designed and implemented by the HR department. At the moment, this discussion is predominately driven by practitioners and focuses on operational implementation. Big data will be a transformative power, but they are shaped by the people in the organization. The HR department can use big data to transform the organization proactively and adapt a new role, or leave this emergent but critical field to other departments. HRM will need to reinvent itself in order to deal with big data and use them for their purposes.

Deficit 4: Big data will force HRM to change and assume a new role in the organization. However, it is unclear what this role will look like.

1.2 State of Research

Big data is the buzzword today and many are willingly jumping onto the bandwagon. Big data are new, ubiquitous, and pervasive. However, big data and their effects on organizations are under-researched. Statements claiming that big data would lead to enhanced objectivism are not entirely true since big data are subjective, never neutral, but contextualized (Johnson 2015). Big data are not capable of knowing everything everywhere and anytime. However, this means that the explanatory power of big data is limited and that there is an inherent data bias within big data that leads to distortion between data and *reality*. Big data may potentially lead to a massive paradigm shift in society (Mayer-Schönberger & Cukier 2013) and especially in research (Puschmann & Burgess 2014). Human interaction and its embeddedness within a social network (or organization), in particular, will be shaped differently through this datafication (Lycett 2013). Data are already everywhere and will increasingly become the general mode of communication. Everything can be transformed into a representation of data (Frankel, & Reid, 2008). Big data will impact social life enduringly. This effect is relatively opaque, however, and differs from context to context (Manovich 2011). There are fragmented discussions about the subjective influence of big data (e.g. Boyd & Crawford 2012, Dalton & Thatcher 2014, Kitchin 2014a, Scholz 2015a, Metcalf & Crawford 2016).

Deficit 5: Big data may be subjective; this subjectivity is discussed in a certain context, but not in organization theory or HRM.

Organizations will be transformed by big data, therefore, becoming complex systems (Scholz 2015b). Furthermore, there is an abundance of influences on an organization, which brings about additional obstacles. An emergent trend is analyzing organizations from a complex systems theory perspective (e.g. Amaral & Uzzi 2007), and from the perspective of dynamization (e.g. Stein & Müller 2012). Organizations are already forced into transformation by external pressure from globalization, but big data will further increase the velocity of such a transformation. Research is currently dealing with several aspects concerning the topic of such transformation, but the digital perspective and, therefore, big data, are still widely neglected. Digitization is undeniably a driver of change (e.g. Castells 2010, Stein 2015), but research is cagey about the topic (Knop 2014).

Deficit 6: Organizations are becoming more complex and dynamic; big data will act as a catalyst for complexity, but research is neglecting it in this context.

Big data will influence the organization, and underestimating their impact will probably be more harmful than dealing with the subject of big data. There are many aspects that make big data interesting. But the most important aspect is that there are big data within an organization that lie fallow. Smart factories and digitization leave a rising pile of data unexploited. No organization that is profit-oriented can look the other way. Big data's role as technological game changer is observable and undeniable. But big data will also change the way we work, although it is quite unclear in what way. It is foreseeable that the amount of data collected will massively increase in the future. Improvements in automation and the development of sensors as well as the gathering of human information will pile up the amount of data collected. In addition to that, data that already exist are normally not forgotten (Rosen 2012), as the capacity of storing data is constantly increasing (Hilbert & López 2011). But what does that mean for people within organizations? Postman singles out two distinct dystopian futures of information (processed data): "Orwell feared those who would deprive us of information. Huxley feared those who would give us so much that we would be reduced to passivity and egoism" (2006: xix). As a result, we are now moving towards a brave new world of data (Scholz, T. M. 2014) in analogy to the title of Huxley (1932). At the moment, both Orwell's and Huxley's predictions appear to be coming true.

Deficit 7: Big data will have an impact on people and organizations, but the potential outcome is still pending and needs further research.

Researchers will deal with big data from a theoretical perspective. Many disciplines discuss big data in very different ways, but thus far lacking a concise theoretical framework. Various existing theories (especially organizational theory) approach big data. Neither, however, is capable of understanding big data entirely. Even whether or not the term big data is precise enough, or whether big data are merely old wine in a new bottle may be debatable. The phenomenon itself, however, will not simply be rationalized away. There is a need to understand big data and the hefty influence of big data on today's world and to utilize this knowledge.

Deficit 8: Big data is not theory-less, however, there are no fitting theories available. This is especially true for organizational theory and HRM.

Up to this point, few publications have dealt with big data in HRM and those few are dominated by practitioners (e.g. Bersin 2012, CIPD 2013, Cornerstone OnDemand 2013, eQuest, 2013, Evolv 2013). In academic research, authors consider looking at individual aspects of big data in HRM, such as the management process (McAfee & Brynjolfsson 2012), analytics (Galagan 2014; Shah et al. 2012), performance (Levenson 2014), talent (Russell & Bennett 2014), workforce management (Miller 2013), and the new employment fields of data scientist (Davenport & Patil 2012, Davenport 2013, 2014), and chief data officer (Lee et al. 2014). There is a gap in the literature with respect to grasping the scope of big data in HRM, scientific discourse is lagging behind practical application (George et al. 2014).

Deficit 9: Big data in HRM are currently driven by practitioners, researchers are already behind them. However, it will be necessary to deal with big data in HRM from a research perspective.

1.3 Terminological Clarification

The term *data* will be omnipresent in the course of this thesis. The plural form will be employed in accordance with the conversation proposed by Kitchin (2014a). He quotes the Oxford English Dictionary:

> In Latin, *data* is the plural of *datum* and, historically and in specialized scientific fields, it is also treated as a plural in English, taking a plural verb as in the *data were* collected and classified.
>
> In modern non-scientific use, however, it is generally not treated as a plural. Instead, it is treated as a mass noun, similar to a word like information which takes a singular verb. Sentences such as *data was* collected over a number of years are now widely accepted in standard English.

While this thesis will refer to data as a plural term, the original version will be retained in quotations. Furthermore, 'big data' when labeling a theoretical concept, will be used in the singular form.

1.4 Objective of the Thesis

Concerning the theoretical foundation of big data, the relationship between human and big data, and the role big data play within an organization, research is currently relatively scarce. Although there are thousands of papers on the subject, many are purely technologically driven and neglect the human aspect of big data. But big data are bound to become an integral part of society and organizations. The human factor (Zuboff 2014) as well as the big data lens on humans (Aiden & Michel 2013) require research and a concise theoretical understanding before actual effects can be analyzed. Big data are not theory-less but, as of yet, lack theory (West 2013, Monroe et al. 2014, Boellstorff 2015). While the obvious perspective on big data is a technological one, big data deeply penetrate the social environment, which is why social knowledge about big data is of upmost importance. Boyd and Crawford accurately describe the current state of research concerning big data as follows: "The era of Big Data has only just begun, but it is already important that we start questioning the assumptions, values, and biases of this new wave of research. As scholars who are invested in the production of knowledge, such interrogations are an essential component of what we do" (Boyd & Crawford 2012: 675).

Nevertheless, the classification of big data within the philosophy of science lags behind. This thesis will be rooted within three distinct philosophies of science. Firstly, in order to capture the impact of big data on society, organizations, and individuals, the thesis attends to the field of science and technology studies to which the relation between society and technology is the object of research. The research stream of organizational theory represents the second philosophy of science, focusing complex systems theory as well as systems theory, cybernetics, and population ecology theory. The third stream is human resource management research. Especially in the context of economic organizations, HRM research focuses on both closing the gap between research and practice, and transforming the organization adequately towards new innovations that emphasize the human factor.

In order to better understand big data and their interrelation with people and, consequently, the role of big data within an organization, a number of goals need to be met:

- **Deriving a theoretical understanding of big data:** It seems that we have a rough understanding of big data. They are vast, however, and there are many definitions.
- **Understanding the impact of big data on the socio-technological, organizational, and human resource-related level:** The technological aspect of big data alone is complex. But the topic of big data becomes even more complex when including society and the individual. Therefore, big data require analysis on different levels. As its main contribution, this thesis elaborates on a theoretical lens on big data from different theoretical perspectives, and constitutes big data as a social construct rather than a technological one.
- **Understanding the effect of big data on any organization, as well as their ability to transform it:** Big data will transform any social system. They will change any economic organization, thus transforming its very structure.
- **Describing the nature of this transformation:** The changes provoked by big data will fundamentally change the role of the human resource department. Therefore, this thesis is not concerned with the possibilities of big data with regards to an organization's employees, but emphasizes the way in which big

data will be employed within the organization, and how the human resource department will perform the task of supervising big data.

Those goals aim towards developing a theoretical model for a data-augmented homeodynamic organization. This model introduces big data into the organization and describes their impact. It will transform the organization in a comprehensive way, enabling it to deal with big data in an efficient way and utilize them to generate a competitive advantage. That is why this thesis will theorize the impact of big data on an organization, on the HR department, and the people within. In the words of Huxley: “I mean, what I feel very strongly is that we mustn’t be caught by surprise by our own advancing technology” (1958).

2. Theoretical Framework

2.1 Big Data

2.1.1 Etymological Origin

The pursuit of understanding big data first requires exploring the term *data*. It subsumes a variety of meanings and ideas. It is loaded with contextual meaning and depends on the beholder's point of view; there are different perspectives of data. That aside, the term can be derived etymologically as follows:

> "English *data* is derived from Latin, where it is the plural of *datum*, which is in turn the past participle of the verb *dare*, "to give," generally translated into English as "something given." Sanskrit *dadāmi* and ancient Greek *δίδωμι* are related forms. While *data* (piece of information) and *datum* (calendar date) are separate lexemes in contemporary English, their association is not accidental; medieval manuscripts frequently closed with the phrase *datum die* (given on ...), effectively time-stamping the preceding text" (Puschmann & Burgess 2014: 1691).

In addition to its variety of context-determined meanings, the denotation of the term has shifted over time. In the 18th century, it represented a rather quantitative understanding as it "was most commonly used to refer to facts in evidence determined by experiment, experience, or collection" (Rosenberg 2013: 33). Rosenberg himself specifies this point of view by claiming that "facts are ontological, evidence is epistemological, data is rhetorical. A datum may also be a fact, just as a fact may be evidence ... When a fact is proven false, it ceases to be a fact. False data is data nonetheless" (2013:18). Nowadays, however, any mention of data is likely to refer to their digital sense. They are perceived as a common resource generated without any effort and without any loss of information. This is a precise description of today's ubiquitous generation of data. It is for this reason that data are often referred to as new oil (Thorp 2012, Helbing 2015) or lead to a new gold rush (Peters 2012). The latest conception of *data* can be outlined as "anything recordable in a relational database in a semantically and pragmatically sound way" (Frické 2015: 652).

Some researchers (e.g. Kitchin 2014a), however, claim that the term *data* fails to precisely capture the described phenomenon in modern contexts. They suggest the use of *capta* (from the Latin word *capere* which means to take) instead. Data in the modern sense are the extraction of elements through observation, recording and other means (Borgmann 2007), *data* or *capta* are taken from all potential data (Kitchin & Dodge 2011). This etymological permutation is described the following way:

> "It is an unfortunate accident of history that the term datum ... rather than captum ... should have come to symbolize the unit-phenomenon in science. For science deals, not with 'that which has been given' by nature to the scientist, but with 'that which has been taken' or selected from nature by the scientist in accordance with his purpose" (Jensen 1952: ix).

Although, the term *capta*, therefore, bears more precision than the term *data*, the term data has become generally accepted.

The logical next step is analysis of the term *big*. Heuristically speaking, *big* describes something large in size, large in number, or involving many people or things. Applying this to data allows for the inclusion of huge data sets and correlates to the challenge of dealing with an "information explosion" (Marron & de Maine 1967: 711) and, subsequently, the belief that this kind of "information overload" (Eppler & Mengis 2004: 325) leads to a "data avalanche" (Miller 2010: 181) or "data deluge" (Bell et al. 2009: 1297), and that we are "facing the waves of big data" (Marder 2015: 2). There is, however, more than meets the eye in the simple term *big data*. In order to draw a more precise picture of the term, it is essential to review its chronological history.

One of the earliest examples of big data analysis is attributed to John Snow in 1854 (Khoury & Ioannidis 2014). London had been struck by an outbreak of cholera, and Snow collected all available data about the deaths and was able to locate their origin to the area around Broad Street. He hypothesized a connection between the outbreak and a specific water pump. Shutting down the pump led to a significant reduction in the number of new infections (e.g. McLeod 2000, Koch 2004, Johnson 2007). Snow collected data, used it to develop a hypothesis and derived an action from it (Khoury & Ioannidis 2014) and this can be described as data-driven science. His results can be seen in Figure 1. There are several other examples that can potentially be retrospectively attributed to the use of big data. Snow's example, however, is exceptionally well documented, ultimately led to the beginnings of geographical epidemiology (Newsom 2006), and is a prominent example of the early visualization of information (Friendly 2008).

Figure 1: Original Map Used by Snow (1854)

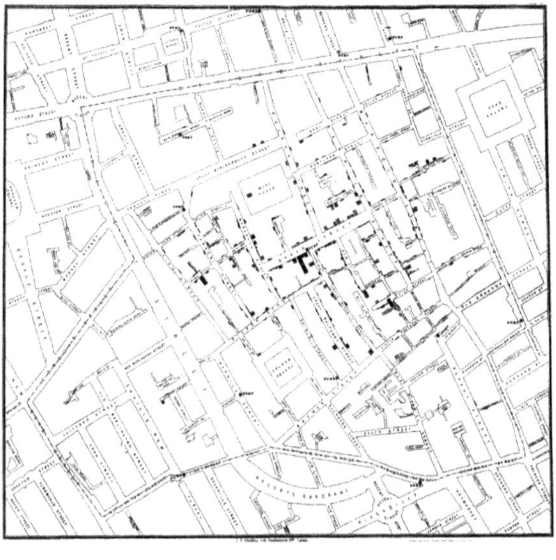

Even though Snow's analytical effort is seen as an example of using big data, the term itself is still relatively young. Its origin, however, is currently under debate. Diebold (2012) attributes the first use of big data to a work by Tilly (1984), and to the use of data analysis for historians. Diebold explains that 'big data' was used in the context of computer science by Weiss and Indurkhya (1998), and by himself in econometrics in 2000. Others (e.g. O'Leary 2013) claim Cox and Ellsworth (1997) and related follow-up research (Bryson et al. 1999) to be the earliest contributions to the term development as used today. Recent research dates the first academic use of 'big data' to 1969 (Scholz 2015a). Even though this early reference uses the term 'big data', the connection to its present conception is vague. Nonetheless, the term 'big data' was used frequently in the 1960s and 1970s. However, it may just be a coincidence that big was combined with data. Table 1 cites several occurrences of the term big data, which clearly foreshadow modern terminology.

Table 1: The Term "Big Data" in the Years 1961–1979

Source	**Quotes with the Term "Big Data"**
U.S. Congress (1961: 197)	"So I think it is quite important that we do not end up doing a *big data*-collecting job, with a quick, casual look at it and that being the end of it."
Kates (1969: 50)	"Most geographers are for *big data* banks, most support an expanded range of census questions, most accept in some vague general way the notion that the more we know about people the better off we are."
DPMA (1970: 8)	"Instead of a *big data* dump where all information collected by government agencies on all Americans would be gathered, he proposes the following..."
Exemplary Miller (1971: 253)	"Eventually, the governance of data centers may fall into the hands of those we now jokingly refer to as 'computerniks', creating a danger that policy will be formulated by information managers who are so entranced with operating sophisticated machines and manipulating *large* masses of *data* that they will not be sufficiently sensitive to privacy considerations."
U.S. Senate (1972: 1270)	"In actual fact, the practice has spawned *big data* center bureaucracies at taxpayer expense. Industry claims millions of dollars are wasted each year – as each federal agency tries to build its own data empire."
Merriam (1974: 40)	"In the future, *big data* storage and retrieval systems will be put into use."
Bassler and Joslin (1976: 300)	"A *big data* center may handle several thousand tapes a day. In addition to tracking the use of tapes and disks, the librarian must be an expert in the care and preservation of the tape and disk media."

Source	Quotes with the Term "Big Data"
Patrick (1977: 35)	"More and more it is becoming apparent that a *big data* processing system requires careful design attention to be given both to the computer processing and the manual processes such as data capture, balancing, error correction, reports distribution that support the computer system."
Müller (1979: 11)	"The dreams of *big data* banks – that would even work – of course raised public fears against the uncontrolled circulation of personal information."

Many issues mentioned in early sources are still current: concerns about data analysis, data accumulation by the government, increasing complexity, and the essential need for the precise design of big data systems. There is also the question of privacy (e.g. Müller 1979). In the book "The assault on privacy", Miller (1971) tackles several aspects of potential surveillance by means of big data (he calls it large data), one of which is the individual loss of control over personal information. Another element is the general tendency to quantify people based on their data. Interestingly, Miller already mentions the delicate challenge this entails for humanity: "Perhaps the single most imperative need at this point in time is a substantial input from human resources to help solve the difficult problem of balancing privacy and efficiency" (1971: 259). While Miller himself was a lawyer, he stressed the crucial need for any discipline to deal with the subject. "There will be no one to blame but ourselves if we then discover that the mantle of policymaking is being worn by those specially trained technicians who have found the time to master the machine and have put it to use for their own purposes" (Miller 1971: 260).

Although a certain interest in big data can be seen, and various people discussed relevant questions (that have yet to be answered), the term 'big data' was used only sporadically in the following years. Apart from Tilly (1984), Cox and Ellsworth (1997), Weiss and Indurkhya (1998), Bryson et al. (1999), and Diebold (2000), no substantial contributions to big data research followed at first. In 2001, however, the paper "3D data management: Controlling data volume, velocity, and variety" by Gartner analyst Douglas Laney moved big data into the focus of business and academia. Laney's article can be understood as the foundation of numerous studies of big data.

2.1.2 Epistemological Conceptualization and Hermeneutical Observations

In order to approximate big data from a hermeneutical perspective, it is necessary to identify the embodiments of data, the first one being its incompleteness:

> "Data harvested through measurement are always a selection from the total sum of all possible data available – what we have chosen to take from all that could potentially be given. As such, data are inherently partial, selective and representative, and the distinguishing criteria used in their capture has consequence" (Kitchin 2014a: 3).

In his seminal work on big data, Kitchin derived various types of data. He categorized them as follows (Kitchin 2014a: 4):

- Form: Qualitative or quantitative
- Structure: Structured, semi-structured, or unstructured
- Source: Captured, derived, exhaust or transient
- Producer: Primary, secondary, or tertiary
- Type: Indexical, attribute, or metadata

Data, in general, are the basis of information and generate knowledge. In the field of knowledge management this process is seen as following a hierarchy (Alavi & Leidner 2001). Data by themselves are informative but do not give insights that are usable in decision making or planning. Only through context (Kidwell et al. 2000), and supplied with meaning and by understanding relationships (Alavi & Leidner 2001), do data become information. Information transforms into knowledge when combined with experience, cognition, and competence (Zins 2007). Knowledge is, therefore, necessary in order to deal with given data and information (Kebede 2010). Other researchers (e.g. Adler 1986, Weinberger 2011) describe the process as a pyramid and an inherent process of distillation that moves up the pyramid, thus reducing complexity, organizing information, interpreting, and finally applying processed data to decisions (McCandless 2010). Weinberger illustrates the process as follows: "Information is to data what wine is to the vineyard: the delicious extract and distillate" (2011: 2). Consequently, data can be processed into something useful, but, thereby, data will be transformed.

As Kitchin notes, "data are never simply just data; how data are conceived and used varies between those who capture, analyse and draw conclusions from them" (2014a: 4). Consequently, data may be sufficiently defined; nowadays however, given their omnipresence data are proverbially multiplying, thus prompting the need for a different framing.

Although the interest in big data coincided with the paper written by Laney (2001), the main reason for the exponential growth in big data can be attributed to the turning point at which storing digital data became cheaper and more cost-effective than storing data on paper (Morris & Truskowski 2003). More and more data are generated digitally and digitization allows them to be shared. Interestingly, many issues affecting the growth of big data follow Moore's law (Schaller 1997) of exponential growth. Some researchers (e.g. Dinov et al. 2014) suggest that although computational capabilities still follow Moore's law, data acquisition behaves according to Kryder's law (Walter 2005), which suggests that data volume is growing at an even higher pace. Computational power, however, remains a main driver for the success of big data, and nowadays it is possible to conduct elaborate big data research on an average computer (Murthy & Bowman 2014).

All things considered, we truly are using *big* data, and the technological perspective suggests that it is growing exponentially. Even though data are ubiquitous in society, however, there is no unified definition of what big data really are. An initial point in the discussion of big data is its classification in dimensions.

This perspective stems from the original paper by Laney (2001), which categorized big data into three dimensions: volume, variety, and velocity. *Volume* denotes the amount of data that is collected. Big data volumes are currently measured in petabytes (1,000 terabytes), however, the amount of data collected is rapidly increasing (McAfee & Brynjolfsson 2012). The dimension of *variety* marks the types and forms in which data are collected. Data can be structured or unstructured, and there are numerous forms: numbers, text, audio, and video (Aakster & Keur 2012), to name only a few. *Velocity* refers to the pace at which data are generated and analyzed. The issue of speed can be dealt with by focusing on data collection, or the challenge of parsing data in real-time (Hendler 2013). Within the course of the following years, a variety of new dimensions were added as depicted in table 2.

Table 2: Dimensions of Big Data

Dimension	Definition
Volume	"E-commerce channels increase the depth/breadth of data available about a transaction (or any point of interaction)" (Laney 2001: 1).
Variety	"Through 2003/04, no greater barrier to effective data management will exist than the variety of incompatible data formats, non-aligned data structures, and inconsistent data semantics" (Laney 2001: 2).
Velocity	"E-commerce has also increased point-of-interaction (POI) speed and, consequently, the pace data used to support interactions and generated by interactions" (Laney 2001: 2).
Veracity	"Data uncertainty. Veracity refers to the level of reliability associated with certain types of data. [...] The need to acknowledge and plan for uncertainty is a dimension of big data that has been introduced as executives seek to better understand the uncertain world around them" (Schroeck et al. 2012: 5).
Variability	"In addition to the speed at which data comes your way, the data flows can be highly variable – with daily, seasonal and event-triggered peak loads that can be challenging to manage" (Troester 2012: 3).
Complexity	"Difficulties dealing with data increase with the expanding universe of data sources and are compounded by the need to link, match and transform data across business entities and systems. Organizations need to understand relationships, such as complex hierarchies and data linkages, among all data" (Troester 2012: 3).

Dimension	Definition
Value	"The economic value of different data varies significantly. Typically there is good information hidden amongst a larger body of non-traditional data; the challenge is identifying what is valuable and then transforming and extracting that data for analysis" (Dijcks 2013: 4).
Viability	"Our first task is to assess the viability of that data because, with so many varieties of data and variables to consider in building an effective predictive model, we want to quickly and cost-effectively test and confirm a particular variable's relevance before investing in the creation of a fully featured model" (Biehn 2013).

Listed here are only the most prominent dimensions used to describe big data. There are, however, several more, including visualization (van Rijmenam 2013), valorization (Özdemir et al. 2013), validity, venue, vocabulary, vagueness (Borne 2014), versatility, volatility, virtuosity, vitality, visionary, vigor, viability, vibrancy, and virility (Uprichard 2013). In addition to using Vs, researchers have recently expanded their choice of dimensions into the Ps, such as privacy (Agrawal et al. 2015), portentous, perverse, personal, productive, partial, practices, predictive, political, provocative, polyvalent, polymorphous, and playful (Lupton 2015).

Academic discourse is currently raising the question of whether to view "big data as merely a shift in scale, reach, and intensity (a quantitative shift) or as a more profound, truly qualitative shift – implying both a shift in being (ontology) and meaning (epistemology)" (Bolin & Schwarz 2015: 2). Bolin and Schwarz (2015) also point out that big data can be classified by a heuristic logic, or can foster a religiously tainted dataism (van Dijck 2014). This may cause an increase in datafied interpretations resulting in a more "anti-hermeneutical impulse (naïve empiricism)" (Bolin & Schwarz 2015: 2). The starting point of the discussion, however, is that "the world grows in complexity, overwhelming us with the data it generates" (Chakrabarti 2009: 32). This means, "it demands a systematic response" (Bowker 2014: 1797).

A great deal of literature on the topic consists of some sort of subliminal judgment of big data. Some sources praise its potential for making the world a better place (e.g. Smolan & Erwitt 2013). According to Pentland, for example, big data will help "build a society that is better at avoiding market crashes, ethnic and religious violence, political stalemates, widespread corruption, and dangerous concentrations of power" (2014: 16). Others focus on the challenges and obstacles that accompany the use of big data, one of which is the assumption that "big data continues to present blind spots and problems of representativeness, precisely because it cannot account for those who participate in the social world in ways that do not register as digital signals" (Crawford et al. 2014: 1667). The two camps both praise and criticize big data in various ways which, although the arguments are versatile, go to show that big data is a multi-faceted term.

Puschmann and Burgess (2014) take a similar approach by conceptualizing big data on the basis of two metaphors. Firstly, they claim that "big data is a force of nature to be controlled" (2014: 1698), which is often associated with the natural force of water. Society is drowning and will deal with data floods or data tsunamis in some way. The authors claim that the analogy of water fits in the sense that water is neutral and able to exist without humans. With the appropriate technology, however, both water and data can be harnessed. The second metaphor is that "big data is nourishment/fuel to be consumed" (2014: 1700) which aligns with the idea of data as "the new oil" (Helbing 2015) and especially the concept of data as a resource. Both of these "metaphors are crucial narrative tools in the popularisation of knowledge" (van Dijk 1998: 22). Nevertheless, the opposing assumptions of the two metaphors strengthen the argument that the term is still nascent.

Big data are highly debatable in terms of the way in which data are analyzed; the term itself seems to be opaque as well. Using dimensions helps approximate the concept of big data, but they only seem to tackle certain aspects, and are, therefore, not sufficient for exhaustively defining big data. Defining big data is a difficult and complex task. In addition, subjective preconceptions influence the process of definition. Statements such as "Across all disciplines, data are considered from a normative, technological viewpoint" (Kitchin 2014a: 12) reveal the obstacles of defining big data in a logical way. Similar to the is-ought problem or Hume's law (Hume 1739), the process of describing big data faces the inherent problem of researchers making statements about what ought to be without being capable of deriving any descriptive statements whatsoever. From a technological viewpoint, a computer cannot differentiate between descriptive and prescriptive. With enough data, anything becomes a standard (Helland 2011). Both human and machine contribute to a big data fallacy that leads to a variety of hermeneutic judgments about big data. From a normative perspective, both camps can be categorized as within subjective objectivity (Leahu et al. 2008) and objective subjectivity (Diller 1997).

If big data had actually marked a scientific revolution (Kuhn 1962) and led to a paradigm shift (Kitchin 2014b) in science as well as in business, former standards would no longer be applicable. Several researchers claim big data as the fourth paradigm by going beyond the experimental, theoretical and computational sciences. Gray states: "The techniques and technologies for such data-intensive science are so different that it is worth distinguishing data-intensive science from computational science as a new, *fourth paradigm* for scientific exploration" (Hey et al. 2009: xix). The question of whether big data lead to a paradigm shift or are merely more 'hype' (Kitchin 2013) or a fad (Marr 2015) may be debatable (e.g. Gandomi & Murtaza 2015). Nevertheless, big data have the ability to disrupt and challenge existing norms and standards (Boyd & Crawford 2012). Such newness and uniqueness, however, would mean that social norms and technological standards are not accurately fitted for such a novel concept as big data.

It is, therefore, essential to incorporate the variety of normative perceptions into an epistemological conceptualization and embrace the hermeneutical bias of researcher and machine. The way in which big data are perceived can be categorized, and there are, in fact, already various categorizations in existence. Kitchin (2014a) divides big data into technical, ethical, political and economic, temporal and spatial, and philosophical viewpoints. De Mauro et al. (2014) classify big data by means of the following four themes: information, technologies, methods, and impact. Boyd and Crawford (2012) categorize big data into cultural, technological, and scholarly phenomena.

As previously mentioned, therefore, the term is opaque, and there are also a variety of definitions available. De Mauro et al. (2015) conducted a survey of existing definitions as shown in table 3. The authors systematically reviewed the literature until on the 3rd of May 2014, their corpus had reached a volume of 1,437 conference papers and articles with the term 'big data' as part of their title or on the list of keywords. The data coincide with the list of definitions postulated by Ward and Barker (2013).

Table 3: Existing Definitions of Big Data

Source	**Definition**	**I**	**T**	**M**	**P**
Beyer & Laney (2012)	High volume, velocity and variety information assets that demand cost-effective, innovative forms of information processing for enhanced insight and decision making.	X		X	X
Dijcks (2012: 3–4)	The four characteristics defining big data are volume, velocity, variety and value.	X			X
Intel (2012: 3)	Complex, unstructured, or large amounts of data.	X			
Suthaharan (2014: 71)	Can be defined using three data characteristics, cardinality, continuity and complexity.	X			
Schroeck et al. (2012: 5)	Big data is a combination of volume, variety, velocity and veracity that creates an opportunity for organizations to gain competitive advantage in today's digitized marketplace.	X			X
NIST (2014: 5)	Extensive data sets, primarily in the characteristics of volume, velocity and/or variety that require a scalable architecture for efficient storage, manipulations, and analysis.	X	X		
Ward & Barker (2013: 2)	The storage and analysis of large and/or complex data sets using a series of techniques including, but not limited to, NoSQL, MapReduce and machine learning.	X	X		

Source	Definition	I	T	M	P
Microsoft (2013)	The process of applying serious computing power, the latest in machine learning and artificial intelligence, to seriously massive and often highly complex sets of information.	X	X	X	
Dumbill (2013: 1)	Data that exceeds the processing capacity of conventional database systems.	X	X		
Fisher et al. (2012: 53)	Data that cannot be handled and processed in a straightforward manner.	X		X	
Shneiderman (2008)	A data set that is too big to fit on a screen.	X			
Manyika et al. (2011: 1)	Data sets whose size is beyond the ability of typical database software tools to capture, store, manage, and analyze.	X	X	X	
Chen et al. (2012: 1166)	The data sets and analytical techniques in applications that are so large and complex that they require advanced and unique data storage, management, analysis, and visualization technologies.	X	X	X	
Boyd & Crawford (2012: 663)	A cultural, technological, and scholarly phenomenon that rests on the interplay of technology, analysis and mythology.		X	X	X
Mayer-Schönberger & Cukier (2013: 12)	Phenomenon that brings three key shifts in the way we analyze information that transform how we understand and organize society: 1. More data, 2. Messier (incomplete) data, 3. Correlation overtakes causality.	X		X	X
Legend: I – Information, T – Technology, M – Methods, P – Impact.					

(adapted from De Mauro et al. 2015: 102)

From the above, De Mauro et al. derived the following definition: "Big data represents the information assets characterized by such a high volume, velocity and variety to require specific technology and analytical methods for its transformation into value" (2015: 103). Beyond this extensive list, there are further definitions of the term 'big data'. Kitchin (2014b: 1–2) characterizes big data as massive in volume, quick in velocity, distinct in variety, exhaustive in its domain, granular in its resolution, relational in its structure, flexible, and scalable in its consistence. Kitchin (2014a) derived four other characteristics from additional literature (Boyd & Crawford 2012, Dodge & Kitchin 2005, Marz & Warren 2012, Mayer-Schönberger & Cukier, 2013):

- "*exhaustive* in scope, striving to capture entire populations or systems (n = all), or at least much larger sample size than would be employed in traditional, small data studies;
- fine-grained in *resolution*, aiming to be as detailed as possible, and uniquely *indexical* identification;
- *relational* in nature, containing common fields that enable the conjoining of different data sets;
- flexible, holding the traits of *extensionality* (can add new fields easily) and *scalable* (can expand in size rapidly)" (Kitchin 2014a: 68).

In addition to these definitions, Dutcher (2014) asked 43 "thought" leaders to give their own definition of big data. Ashlock (Chief Architect of Data.Gov), for example, defines big data the following way:

> "While the use of the term is quite nebulous and is often co-opted for other purposes, I've understood 'big data' to be about analysis for data that's really messy or where you don't know the right questions or queries to make – analysis that can help you find patterns, anomalies, or new structures amidst otherwise chaotic or complex data points."

Many definitions converge towards those suggested by Kitchin. Upadhyay (CEO of Lattice Engines), however, implies that big data may be an umbrella term bearing diverse meanings. O'Neil (Columbia University) attributes rhetorical potential to big data, thus identifying it as a tool of manipulation. To Murphy (Consulting Data Scientist), the word *big* is the key. His qualitative evaluation of the term 'big data' emphasizes its complexity of definition. *Big* data are more than meets the eye.

Due to the fact that big data are too wide-ranging and vague in definitions, some practitioners declare big data as already dead (e.g. de Goes 2013). Although big data may be vague and cannot be precisely pinpointed, the term itself describes the current challenge of datafication as faced by society, organizations, and individuals alike. A precise definition of big data is, therefore, probably not possible, as such a definition would turn out to be *big* as well. Jacobs, however, suggests the following meta-definition, picking up the umbrella concept of Upadhyay: Big data refer to "data whose size forces us to look beyond the tried-and-true methods that are prevalent at that time" (2009: 44). Subsequently, "The challenge is not just a technological one: the selection, control, validation, ownership and use of data in society is closely related to social, ethical, philosophical and economic values of society" (Child et al. 2014: 818). Although Anderson proclaimed the end of theory in asserting that "with enough data, the numbers speak for themselves" (Anderson, 2008), big data seem to, thus far, have fostered theory-building (Boellstorff 2015, Tohki & Rauh 2015). In addition to the possibility of asking questions that could not be asked before (and obtaining different answers) (Weinberger 2013, Hand 2016), theoretically untangling the construct of big data is a Herculean task in itself. Big data have long been entangled with society, and we need to cope with that (Floridi 2012) – a discourse that is necessary (Rayport 2011, Barabási 2013), especially considering the omnipresence of big data.

Big data are already living a life of their own. Big data also connect to different flourishing concepts already in existence. This thesis aims at the delimitation of big data from data mining, its connection to algorithms, machine learning, and artificial intelligence. All above concepts are often used synonymously, and are closely intertwined with big data.

2.1.3 Delimitation from Related Terms

2.1.3.1 Data Mining

Big data are often associated with data mining and sometimes both terms are incorrectly used interchangeably. Although both terms deal with data, there are significant differences. Data mining is the computational process of discovering patterns in (large) data sets (Han et al. 2012). The process is often described as knowledge discovery in databases (Fayyad et al. 1996): the use of algorithms, statistical tools, and machine learning in order to extract patterns previously unknown. By identifying clusters, detecting anomalies, locating dependencies, and finding correlations, data mining supports the process of data analysis (Larose 2014).

Data mining is a method or a tool used in handling (big) data. Using data mining merely reveals patterns, and represents only *one* step in the process of knowledge discovery (Fayyad et al. 1996). This embeddedness of data mining into a *bigger* process can be integrated in the steps of knowledge discovery in databases, as proposed by Fayyad et al. (1996: 41):

- Selection
- Preprocessing
- Transformation
- Data mining
- Interpretation/Evaluation
- Knowledge

Data mining means scouring a haystack of data in order to find something other than hay, and maybe the metaphorical needle. It is the process of searching around *existing* data sets and finding information or knowledge that has thus far been unknown. The patterns and signals that may be hidden amidst the noise are what such a system is mining for. Data mining, therefore, does not serve the purpose of collecting data, it merely uses available data. Selecting data and interpreting data does not lie within the realm of data mining. Researchers, thus, point out that the term 'data mining' fails to sufficiently describe the actual process. Han et al. (2012: 6) coin a more suitable term: "knowledge mining from data". Other terms for data mining are knowledge extraction, data/pattern analysis, and data archeology.

Data mining is related to the concept of uncovering patterns without devising preliminary hypotheses (Scholz & Josephy 1984). This explains researchers' tendencies to talk about and deal with data dredging (Selvin & Stuart 1966), data snooping (Sullivan et al. 1999), or spurious correlations (Jackson & Somers 1991). It is crucial, however, to emphasize that data mining can be part of a big data analysis, even though it marks only one step in the analysis. Data mining on its own leads to several statistical (Hand 1998), as well as ethical issues (Seltzer 2005). Many of those issues, such as overfitting (Elkan 2001), could be tackled by integrating data mining into a holistic big data value chain (Miller & Mork 2013).

2.1.3.2 Algorithms and Machine Learning

In order to describe algorithms, it is essential to understand their relevance. Cormen et al. (2009: xii) clarify that "before there were computers, there were algorithms. But now that there are computers, there are even more algorithms, and algorithms lie at the heart of computing". Due to the exponential growth of digitization and the abundance of data, those algorithms have become more relevant, increasing their influence on society. Beer attributes to algorithms "the capacity to shape social and cultural formations and impact directly on individual lives" (2009: 994). Algorithms influence everybody's everyday life (Pasquale, 2015). Some researchers foresee a future of algorithms "running the world" (Lisi 2015: 23). An algorithm can be defined as follows:

> "Informally, an *algorithm* is any well-defined computational procedure that takes some value, or set of values, as *input* and produces some value, or set of values, as *output*. An algorithm is thus a sequence of computational steps that transform the input into the output" (Cormen et al. 2009: 5).

The connection with big data is evident, as data constitute an algorithm's input and are generated as its output. This makes algorithms a tool to transform data. On the basis of explicit instructions, a computational device follows an algorithm step by step, with a finite amount of input (Boolos & Jeffrey 1974).

The line between algorithms and machine learning is relatively hazy and sparks the question: "Can machines think?" (Turing 1950: 1). Although Turing states that a machine probably cannot think, he suggests that we consider its potential to learn.

> "A computer program is said to learn from experience E with respect to some class of tasks T and performance measure P if its performance at tasks in T, as measured by P, improves with experience E" (Mitchell 1997: 2).

As opposed to humans, a computer program may learn from errors or learn by executing tasks repetitively, but will in fact have "no idea what it's doing" (Schank 2015: 132). Algorithms are incapable of learning how to learn (Argyris & Schön 1996). A machine's learning process can be allocated to two distinct learning scenarios (Mohri et al. 2012). *Supervised learning*, during which a teacher (human) teaches a student (machine) new things, means that training data is available to reveal the instances in which input and output are correctly connected. *Unsupervised learning*, on the other hand, has no sample output data, forcing an algorithm to search for patterns, correlations, or clusters to discover similarities. Such a definition has similarities to that of data mining, despite data mining being more static and using a finite set of data as well as a strict method of mining. Another type of learning is through the method of reinforcement learning (Sutton & Barto 1998). When learning through reinforcement, there is no knowledge about the correct output. Measuring the correctness of said output, however, becomes possible in interaction with, and through feedback from, the environment.

In his seminal work, Minsky (1961) categorized existing problems for algorithms as search, pattern-recognition, learning, planning, and induction. Minsky reviewed these methods and concluded that they still display many inefficiencies. Kosko (2015) claimed that not much has changed from an algorithmic perspective. He explains that people are still using algorithms that are decades old. Most of the progress achieved in recent years can be attributed to the increase in computational power, as well as to the increase in the amount of data being processed. Similar to the data mining delimitation, algorithms or machines are "not thinking, nor anything like thinking" (Schank 2015: 132). They are, however, now capable of analyzing huge piles of data, albeit still fairly inefficiently, as noted by Minsky (1961). It seems that in the early development of computational algorithms, elements such as speed, compactness, and elegance (Knuth 2011) were admired, while nowadays the use of brute-force (Fellows et al. 2012) and number-crunching (Vaux 2013, Schank 2015) may suffice due to the exponential increase in computational power. Kosko proposes that one of the most prominent algorithms in unsupervised learning is still k-means clustering (in MacQueen 1967). Even though it now carries diverse names, the general idea behind the algorithm remains unchanged. In the context of supervised learning, a popular algorithm is backpropagation (Rumelhart et al. 1986). Kosko concludes that the future means "old algorithms running on faster computers" (2015: 426).

This development may point towards the foreseeable future use of algorithms. Big data, however, have affected the potential of algorithms dramatically. Although most algorithms are based on a simple logic, they exhibit a tendency to become more complex; so complex and advanced, in fact, that the outcome is inscrutable and incomprehensible, even for the engineers behind the respective algorithm (LaFrance 2015). Although this may be bearable, there are several aspects that reveal an underlying problem. (1) People rely *overconfidently* on data (Miller, C. C. 2015). (2) Due to the opaqueness of data, everything that follows an organizing principle may be misinterpreted as transparency (Gillespie 2012). (3) Algorithms maximize myopia due to their focus on solving problems in short- or even real-time (Luca et al. 2016). (4) Algorithms are designed by people. They could implement anything they want in an algorithm and, as stated in (1), understanding these algorithms is becoming increasingly difficult. It is possible to include a certain ideology (Maher 2012) in an algorithm, or opportunities to commit fraud (Parameswaran 2013), and loopholes that allow for "gaming the system" (Rieley 2000).

2.1.3.3 *Artificial Intelligence*

There are counter-positions to the statement by Kosko and the halt of algorithmic development. Valiant (1984, 2013), for example, proposes the framework of, probably approximately, correct learning. This understanding of the learning process implies that computers may be capable of acquiring knowledge in a similar way to humans and, therefore, "in the absence of explicit programming" (Valiant 1984: 1134). This kind of ability is essential for improving the work between teacher (human) and student (machine) "where humans may be willing to go to great lengths to convey their skills to machines but are frustrated by their inability to articulate the algorithms they themselves use in the practice of the skills" (Valiant 1984: 1142). Another emerging field is deep learning (Arel et al. 2010), as a type of learning inspired by neural networks (Cheng & Titterington 1994). Deep learning deals with "multiple [...] layers of nonlinear information processing and methods for supervised or unsupervised learning [...] at successively higher, more abstract layers (Deng & Yu 2013: 201). Consequently, improvements in machine learning are strongly connected to the research and development of *artificial intelligence* (Deng & Yu: 2013). The development of artificial intelligence was predicted by Turing (1950: 8): "I believe that at the end of the century the use of words and general educated opinion will have altered so much that one will be able to speak of machines thinking without expecting to be contradicted." The statement may be bold, yet we are already surrounded by a variety of artificial intelligences (e.g. Siri or Cortana).

Generally speaking, artificial intelligence can be defined as the "design of intelligent agents" (Poole et al. 1998: 1). This is a general definition of a broad term, but marks a significant difference to fields concerned with the human mind, like psychology: researching artificial intelligence does not merely involve understanding, but also building artificial intelligences (Russel & Norvig 1995). For that reason, an abundance of more precise definitions can be sorted into the following categories:

- Systems that think like humans
- Systems that think rationally
- Systems that act like humans
- Systems that act rationally (Russel & Norvig 1995: 5)

The two categories linked to humans reveal a difficult bottleneck. The reason for this is that "We don't have sufficient ability to observe ourselves or others to understand directly how our intellects work" (McCarthy 2007: 1175). As a result, such definitions are often used when approaching the issue from a theoretical perspective.

Artificial intelligence can, therefore, currently be codified into three types, the first being *weak* (Searle 1980) or *narrow AI* (Hutter 2009). Intelligence of this type specializes in one specific area and is only capable of performing well in this field. The chess machine Deep Blue, while capable of beating humans at chess, is not able to do much else. In today's world, we are surrounded by such AI, for example spam filters, Google Translate, autopilot, self-driving cars, and so on. The second type of AI is the *strong artificial intelligence* (Searle 1980), sometimes called *full* (Bainbridge

2006), *human-level* (Nilsson 2005), or artificial *general* intelligence (Voss 2007). This type of AI refers to the idea that such an intelligence could perform any task. A machine of that kind is intelligent beyond a narrow spectrum and no longer specialized in only one specific field, thus capable of many tasks. Such an AI would pass the Turing Test (Turing 1950) and the Chinese Room Test (Searle 1980) and would be indistinguishable from a human. Finally, there is the third type of *superintelligence*. As yet, this type of AI is merely a theoretical mind game and a popular theme in science fiction. The very idea, in fact, connects artificial intelligence and its potential for human extinction (Barrat 2013). Bostrom defines it the following way:

> "By a 'superintelligence' we mean an intellect that is much smarter than the best human brains in practically every field, including scientific creativity, general wisdom and social skills. This definition leaves open how the superintelligence is implemented: it could be a digital computer, an ensemble of networked computers, cultured cortical tissue or what have you. It also leaves open whether the superintelligence is conscious and has subjective experiences" (Bostrom 2006: 11).

In addition to potential dystopian consequences, this type of artificial intelligence sparks an interesting discussion: would we even be able to understand an artificial superintelligence? In today's world, artificial intelligence predominantly mimics intelligence (Munakata 1998) without actually understanding (Hearst 2015). As McCarthy states, "Much of the public recognition of AI has been for programs with *a little bit of AI and a lot of computing*" (2007: 1175). Consequently, the development of artificial intelligence has not come as far as some people believe (Dreyfus 1965, Hopgood 2003, Epstein 2015). Others have identified AI as a threat to humanity (Hawking et al. 2014). Artificial intelligence is still predominantly human (de Biase 2015), due to the fact that AI is initially programmed by humans. For that reason, Dobelli (2015) refers to artificial intelligence as *humanoid thinking*. Any artificial intelligence will be restricted by a certain humanoid framework. There may also be the possibility of *alien thinking* of sorts. Such thinking will differ greatly from anything we know. This type of thinking, however, requires its own evolutionary path, "not just evolutionary algorithms" (Dobelli 2015: 99). Alien thinking, while imaginable, will therefore need some time to evolve. Artificial intelligence may be seen as something alien when humans no longer understand the underlying algorithms. For the sake of a clearer distinction, Kosslyn (2015) recognized the difference as close AI and far AI. Clark (2015) replies that, although artificial intelligence exposed to big data and deep learning will lead to knowledge that seems opaque, it will "end up thinking in ways recognizably human" (Clark 2015: 156). As a result, current artificial intelligence depends heavily on big data, without which many current systems would perform insufficiently. Any current self-driving car depends on data, be it generated by the car itself through sensors, or data from other sources. Google Translate is an example of big data rather than a weak AI system.

2.1.4 Big Data Pitfalls

In today's society, many people attempt to describe big data and outline their potential. Big data have been placed on a pedestal as being something unique and precious (Mayer-Schönberger & Cukier 2013). Today's discourse of big data makes it appear to be the solution to all problems of society (Steadman 2013), and capable of making the world a safer and better place (Olavsrud 2014). Inherent in this discourse is the belief that making something more data-based or data-driven will lead to more objective and considered decisions (McAfee & Brynjolfsson 2012), however, generating and using big data is a process that is not as clean or sterile as some researchers suggest. Many believe in "objective quantification" (van Dijck, 2014: 198), but big data are "messy" (Harford 2014: 14) and even just the collection of data causes a manipulation or "preconfiguration of data" (van Dijck & Poell 2013: 10). For that reason, ascribing an "aura of truth, objectivity, and accuracy" to big data would be a fallacy (Boyd & Crawford 2012: 664). Nevertheless, literature links big data to a variety of pitfalls. In the related literature, many researchers often refer to three papers (Boyd & Crawford 2012, Richards & King 2013, Dalton & Thatcher 2014). These papers present a systematization of obstacles in the usage of big data. In the following I will present them and explain why I chose Boyd and Crawford's (2012) systematization.

In their 2012 article, Boyd and Crawford develop six aspects of potential pitfalls of big data:

1. Big data change the definition of knowledge
2. Claims of objectivity and accuracy are misleading
3. Bigger data are not always better data
4. Taken out of context, big data lose their meaning
5. Accessibility does not make them ethical
6. Limited access to big data creates new digital divides

Dalton and Thatcher (2014) postulate a different systematization. They do, however, discuss big data at a more philosophical level and contribute to the meta-level discourse about big data. Big data are a highly social entity, which is why the authors focus on the lack of objectivity in big data:

1. Situating 'big data' in time and space
2. Technology is never as neutral as it appears
3. 'Big data' do not determine social forms: confronting hard technological determinism
4. Data are never raw
5. Big isn't everything
6. Counter-data exist

On the other hand, Richards and King (2013) deconstruct big data by explaining that big data always come with a tradeoff. Although big data contribute to transparency, identity and power equality in certain ways, big data increase opaqueness, anonymity and power inequality in other ways. Richards and King, therefore, postulate the following paradoxes:

1. The transparency paradox
2. The identity paradox
3. The power paradox

There are several other papers (e.g. Mittelstadt & Floridi 2015, Saqib et al. 2015, Hilbert 2016) that tackle challenges related to big data, but most are relatively congruent and point out similar pitfalls.

In the following description of big data pitfalls, I will use the six aspects introduced by Boyd and Crawford (2012), because they are broader than the paradoxes by Richards and King (2013) and more precise than the systematization given by Dalton and Thatcher (2014). Boyd and Crawford (2012) seem to cover all relevant pitfalls of big data and, furthermore, they are giving a precise description of those pitfalls.

2.1.4.1 Big Data Change the Definition of Knowledge

Boyd and Crawford (2012) claim that big data will fundamentally change the way we view the working world and the production process. They compare the situation to Fordism (Amin 1994), which is closely linked to the theory of Taylorism (Taylor 1911), which dehumanized work in the early 20th century and only focused on the mechanic parts and automation involved in the work process. The changes that accompanied mass production extensively transformed the relationship between work and society. Computerization and digitization (Zuboff 1988, 2014) are also currently in the process of changing this relationship radically, and trends like automation may have ground-breaking consequences for the labor market (Frey & Osborne 2013).

Big data may contribute to those changes (Davenport 2014), but predominantly change the way people think. Boyd and Crawford (2010: 153) cite Latour as follows: "Change the instruments, and you will change the entire social theory that goes with them." Puschmann and Burgess figuratively describe this change: "before, we were starved for data; now we are drowning in it" (2014: 8). The abundance of data has changed the perception of knowledge drastically. Big data allow for researchers and practitioners to access information in real-time, and enable them "to collect and analyse data with an unprecedented breadth and depth and scale" (Lazer et al. 2009: 722). Big data have led to a supposedly epistemological paradigm shift (Puschmann & Burgess 2014). The computational turn in knowledge (Thatcher 2014) and the proposed fourth paradigm of data-intensive science discovery (Hey et al. 2009) have led to an environment in which "gather data first, produce hypotheses later" (Servick 2015: 493) represents an approved research approach. Kitchin describes these new data analytics as "seek(ing) to gain insights 'born from the data'" (2014b: 2). Following this argument, Anderson (2008) almost polemically proclaimed the end of theory:

he ability to cheaply and easily gather large amounts of data does have advantages: mple sizes can be larger, testing of theories can be better, there can be continuous sessment, and so on. But data-driven science, the 'fourth paradigm', is a chimera. ience needs problems, thoughts, theories, and designed experiments. If anything, ience needs more theories and less data" (Frické 2015: 660).

sequently, big data can be seen as requiring a meta-theory which consists of ous theories and is, therefore, steeped in theories and methods (Mayer-Schön-er 2014). Big data are both "theory-laden" and "tainted by theory" (Frické 2014: Big data lead to an abundance of data and, therefore, more information. This hat may make distilling knowledge from big data substantially more difficult bs 2009), and calls for a focus on scientific rationale of methodological rigor hi & Rauh 2015) and awareness (Ruths & Pfeffer 2014), two variables more gly demanded than ever before (Lazer et al. 2014).

.2 *Claims of Objectivity and Accuracy Are Misleading*

basis of numerous approaches favoring big data is the underlying claim that, due e mere abundance of data, big data are highly objective (Lukoianova & Rubin) and, due to the variety of sources, more accurate (McAfee & Brynjolfsson). People believe overconfidently that data-driven decisions are superior due eir objectivity (Miller, C. C. 2015).

decision made by the numbers (or by explicit rules of some other sort) has at least e appearance of being fair and impersonal. Scientific objectivity thus provides an nswer to a moral demand for impartiality and fairness" (Porter 1996: 8).

claim makes sense at first sight, but big data are, in fact, highly subjective (Dal-& Thatcher 2014). Some, however, attribute objectivity to big data as a result of nical dispositions. After all, data are collected from sensors, saved into log-files, processed by computers. Their objectivity, thus, results from their mechanical re. For the sake of the argument, let us *assume* there is such a thing as *truly ctive* big data. The critical problem here is that big data are not self-explanatory lier 2010). There is an essential need for somebody to interpret this certain set in order to extract any knowledge from it. Said interpretation could be by man or by a machine. Human and machine often collaborate in some way, but party always leads the interpretation. For that reason, any interpretation is enced by either a human or a machine. A set of big data will be manipulated transformed in one way or another, which renders it less than objective (Dal-& Thatcher 2014). Bollier questions the objective truth of big data, because any action with data will lead to subjective contamination: "Can the data represent bjective truth' or is any interpretation necessarily biased by some subjective or the way that data is 'cleaned'?" (2010: 13). Therefore, Metcalf and Crawford 6) call for a theory on data subjectivity.

"This is a world where massive amounts of data and applied mathem other tool that might be brought to bear. Out with every theory c from linguistics to sociology. Forget taxonomy, ontology, and psych why people do what they do? The point is they do it, and we can it with unprecedented fidelity. With enough data, the numbers spe (Anderson 2008).

Consequently, there is no need to construct hypotheses or cc (Prensky 2009). Formerly unknown patterns will be discovered other words, correlation trumps causation (Lycett 2013). Kitch argumentation the following way:

1. "Big Data can capture the whole of a domain and provide fu
2. there is no need for a priori theory, models or hypotheses;
3. through the application of agnostic data analytics, the data selves free of human bias or framing, and that any pattern within Big Data are inherently meaningful and truthful;
4. meaning transcends context or domain-specific knowledge, th ed by anyone who can decode a statistic or data visualization

Such a description prompts a scientific déjà vu of the statistical b and positivism. Comte's formula in particular, "Savoir pour pr pouvoir" (Comte cited in Merton 1936: 898), translated into "t predict, to predict in order to control" (Clarke 1981: 90), can be li concerning big data. Kitchin, however, characterizes this particul "fallacious thinking" (Kitchin 2014b: 4) and points out the limita decisions (Lohr 2012). Frické (2014) used the words of Popper (that although data can now be collected easily and in a more gr ever before, big data will not translate into useful observations b

"The belief that we can start with pure observations alone, without a ture of a theory, is absurd; as may be illustrated by the story of the n his life to natural science, wrote down everything he could observ his priceless collection of observations to the Royal Society as induc story should show us that though beetles may profitably be colle may not" (Popper 1963: 478).

Claims such as that there is "no need for a priori theory" are bein for example by Frické (2014) who defines data as, per se, enta assumptions of sorts. Knowledge is generated by means of big d and there may be an end to a certain type of theory, but big da new types of theories (Boellstorff 2015, Tokhi & Rauh 2015). C eses and then collecting data may no longer be feasible, since are already available (Frické 2015). In the conclusion of his pap Andersons's claim, demanding focus on thorough and precise sc than deleting theory and using only big data:

Co
var
ber
652
is v
(Ja
(To
stro

2.1

The
to t
201
201
to t

Thi
ton
tec
and
nat
obj
(Bo
dat
a h
one
inf
and
ton
int
an
filt
(20

A closer look at the human interpreter reveals several interpretation biases. "Data are perceived and interpreted in terms of the individual perceiver's own needs, own connotations, own personality, own previously formed cognitive patterns" (Krech & Crutchfield 1948: 94). The following statements are similar arguments: "Disciplines operate according to shared norms" (Gitelman & Jackson 2013: 7). "Individuals construct their own subjective social reality based on their perception of the input" (Bless et al. 2004: 2). In a recent experiment, Silberzahn and Uhlmann (2015) gave 29 research teams the same data set and let them analyze it. Interestingly, the results varied greatly and the authors attributed this variance to the fact that "any single team's results are strongly influenced by subjective choices during the analysis phase" (Silberzahn & Uhlmann 2015: 191). Although it is unclear what influenced the researchers, whether it was their own preconceptions or the data set itself (Griffiths 2015), it seems as though the process of data interpretation itself is "inherently subjective" (Boyd & Crawford 2012: 667). Decisions are based on subjective perception and, therefore, deviate from rational decisions (Tversky & Kahneman 1974). Such biases (Kahneman & Tversky 1973) permeate any decision and, therefore, will influence data interpretation. Arnott (2006) identified more than 37 cognitive biases in the literature and Yudkowsky (2008) reports twelve cognitive biases (only five overlaps). Consequently, the lists are less than comprehensive and, especially in the context of data interpretation, far from exhaustive. For Table 4, I selected the most relevant biases from the work of Arnott (2006) and Yudkowsky (2008) and added further biases that fit the context of data interpretation. These authors used different sources for the definitions of the types of biases, due to the reason that these definitions are more precise and more concise.

Table 4: Selection of Cognitive Biases

Types of Bias	Definition
Hindsight Bias (Fischhoff & Beyth 1975)	"... refers to people's tendency to alter their perception of the inevitability of an event once they know the outcome of the event" (Christensen-Szalanski & Willham 1991: 147).
Correlation Bias (Tversky & Kahneman 1973)	"The subjects markedly overestimated the frequency of co-occurrence of natural associates, such as suspiciousness and peculiar eyes. [...] In their erroneous judgments of the data to which they had been exposed" (Tversky & Kahneman 1974: 1128).
Confirmation Bias (Wason 1960)	"... means that information is searched for, interpreted, and remembered in such a way that it systematically impedes the possibility that the hypothesis is rejected – that is, it fosters the immunity of the hypothesis" (Oswald & Grosjean 2004: 79).

Types of Bias	Definition
Overconfidence Bias (Adams & Adams 1960)	There are "3 distinct ways in which the research literature has defined overconfidence: (a) overestimation of one's actual performance, (b) overplacement of one's performance relative to others, and (c) excessive precision in one's beliefs" (Moore & Healy 2008: 502).
Apophenia (Conrad 1958)	"... the tendency to find meaningful patterns in meaningless noise" (Shermer 2008).
Base Rate Fallacy (Meehl & Rosen 1955)	"The base-rate fallacy is people's tendency to ignore base rates in favor of, e.g., individuating information (when such is available), rather than integrate the two" (Bar-Hillel 1980).
Naïve realism (Ross & Ward, 1996)	"People think, or simply assume without giving the matter any thought at all that their own take on the world enjoys particular authenticity and will be shared by other open-minded perceivers and seekers of truth" (Pronin et al. 2002: 369).

(on the Basis of the Lists of Biases by Arnott 2006: 60–61 and Yudkowsky 2008: 91–119)

Although this is only a sample of potential cognitive biases, those biases reveal the subjectivity of the interpreter. There may be several ways to mitigate those cognitive biases in order to achieve more rational decisions (Burke 2007), but a machine might be far superior to a human in terms of subjectivity, especially since a machine, algorithm, or artificial intelligence, is supposedly more rational than a human. A machine may be a superior interpreter. This is, however, not the case, since "there is no automatic technique for turning correlation into causation" (Spiegelhalter 2014: 264).

Statistics are, generally speaking, a process that takes place post hoc (Frické 2015). There are several statistical biases that potentially distort any data set in one direction or another. One strong distortion factor can be attributed to the p-values in statistical findings. Significant results are essential for publishing research findings (Vidgen & Yasseri 2016).

> "P-values are used in Null Hypothesis Significance Testing (NHST) to decide whether to accept or reject a null hypothesis (that there is no underlying relationship [...] between two variables)" (Vidgen & Yasseri 2016: 1).

Common understanding is that p-values lower than .05 are sufficient to classify findings as statistically significant (Ioannidis 2005). Taking that into account, there is no complete certainty as to whether or not hypotheses on the basis of statistical data will be rejected. There is a chance that a hypothesis could be falsely rejected (type I error) or falsely not rejected (type II error), as shown in Table 5.

Table 5: Type I Errors and Type II Errors

		Null Hypothesis (H0) is	
		True	False
Judgment of Null Hypothesis (H0)	Reject	Type I Error (False Positive)	Correct Inference (True Positive)
	Fail to Reject	Correct Inference (True Negative)	Type II Error (False Negative)

(Sheskin 2004: 54)

Although the threshold value of .05 seems rigorous, there is a chance that out of 100 independent tests, approximately five tests are either false positives or false negatives (Frické 2015). Such an example simplifies the problem too drastically. The type of error, however, occurs commonly in statistical findings (Ioannidis 2005). Although data "speak for themselves" (Anderson 2008), the results derived are erroneous. Even the most rigorous data analyses are susceptible to type I and type II errors, and these errors have consequences, as Ioannidis (2005: 696), well documented, describes: "It can be proven that most claimed research findings are false." According to this, we discover potential patterns where there are none and we overlook others (Shermer 2008). Those correlations can be attributed to logical and comprehensible explanations and lead to spurious correlations (Jackson & Somers 1991). Big data intensify those problems due to their nature, in the sense that there are many observations (n) and an even larger number of parameters (p) which aggravate these obstacles.

> "Big Data means that we can get more precise answers; this is what Bernoulli proved when he showed how the variability in an estimate goes down as the sample size increases. But this apparent precision will delude us if issues such as selection bias, regression to the mean, multiple testing, and overinterpretation of associations as causation are not properly taken into account. As data sets get larger, these problems get worse, because the complexity and number of potential false findings grow exponentially. Serious statistical skill is required to avoid being misled" (Spiegelhalter 2014: 265).

There are also several other error sources, such as sub-setting, overfitting, stepwise regression, univariate screening, and dichotomizing continuous variables that could lead to data-driven "hornswoggling" (Frické 2015: 657).

> "'Correlation is enough'. We can stop looking for models. We can analyze the data without hypotheses about what it might show. We can throw the numbers into the biggest computing clusters the world has ever seen and let statistical algorithms find patterns where science cannot... With enough data, the numbers speak for themselves" (Anderson 2008).

Consequently, Anderson's proposition is wrong.

Finally, data interpretation seems to be highly subjective, regardless of the type of interpreter. The interpreter, however, is not the only source of subjectivity. At this point, the initial claim of truly objective data proves to be non-maintainable. Even with more data, granular observation, and further parameters, sampling biases remain inherent to any set of data (Crawford, 2013). "Indeed, all data provide oligoptic views of the world: views from certain vantage points, using particular tools, rather than an all-seeing, infallible God's eye view" (Kitchin 2014b: 4). In conclusion, any researcher "can be too influenced by preconceptions, or too influenced by the data" (Griffiths 2015: 141). Any type of data, however, is subjective and big data especially are often collected with a certain intent and often repurposed for different goals (Schneier 2015). Consequently, from the point of data creation onwards, any data exhibit a hereditary subjectivity, and removing this distortion resolves around a "careful application of statistical science" (Spiegelhalter 2014: 265).

2.1.4.3 Bigger Data Are Not Always Better Data

Another claim commonly propagated in the field of big data research is that big data are interchangeable with whole data (Boyd, & Crawford, 2012), or that big data are *n = all* (Amoore & Piotukh 2015). Mayer-Schönberger told Harford (2014: 17) in a personal discussion the following:

> "A big data set is one where "n = all" – where we no longer have to sample, but we have the entire background population. [...] And when "N = All" there is indeed no issue of sampling bias because the sample includes everyone."

Claiming n = all seems like an audacious assumption and many researchers reject the thesis of big data as not being merely a sample but being a complete set (e.g. Boyd & Crawford 2012, Bowker 2014, Lagoze 2014, Amoore & Piotukh 2015, Tokhi & Rauh 2015, Hand 2016). Boyd and Crawford (2012) refute this assumption using the example of Twitter, which is often used as a source representative of "all people" (2012: 669). In accordance with this argument, Leonelli notes that "having a lot of data is not the same as having all of them; and cultivating such a vision of completeness is a very risky and potentially misleading strategy" (Leonelli 2014: 7).

Following the analogy that big data are n = all, humans are, metaphorically speaking, surrounded by data. "The world and its universe are, to anything or anyone with senses, incomprehensibly big data" (Andrejevic 2014: 1675). It is however not possible for anybody to access all the data. An intelligence of that kind would be comparable to Laplace's daemon:

> "We may regard the present state of the universe as the effect of its past and the cause of its future. An intellect which at a certain moment would know all forces that set nature in motion, and all positions of all items of which nature is composed, if this intellect were also vast enough to submit these data to analysis, it would embrace in a single formula the movements of the greatest bodies of the universe and those of the tiniest atom; for such an intellect nothing would be uncertain and the future just like the past would be present before its eyes" (Laplace 1951: 4).

Figure 2: Conceptual Evolution of Data over Time (Scholz 2015a)

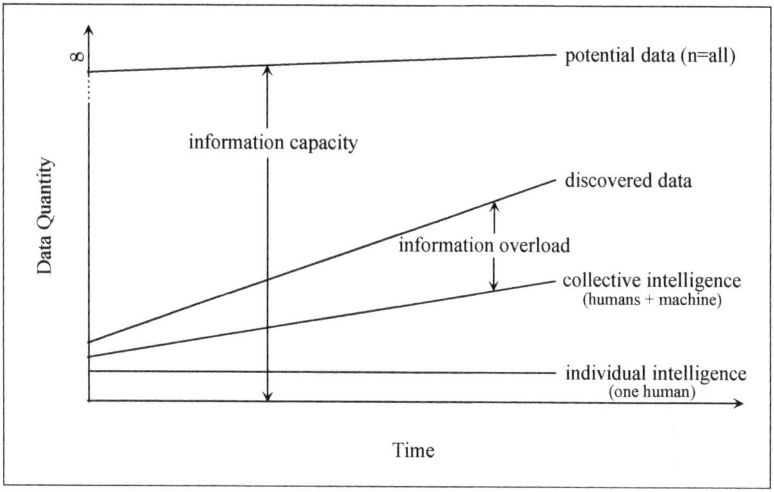

As shown in Figure 2, data are discovered at staggering speed (Miller 2010). Data are also becoming more granular, more singular, and more heterogeneous (Kucklick 2014). Any individual would be overburdened by this deluge of data (Anderson 2008) and could only grasp a small portion of all data. Data absorption can be increased by means of collaboration with other humans, as well as with machines. This is often referred to as collective intelligence (Bonabeau 2009). Deriving from the former statement that computational power is growing according to Moore's law and the acquisition of data, here data discovery follows Kryder's law: the discrepancy between discovered data and collective intelligence has been increasing. This information overload (Eppler & Mengis 2004) cannot be conquered even with the help of technology. Following the description of wholeness of potential data, it seems logical to assume that potential data cannot be exhausted at any time. This assumption becomes clear in the phenomenon that is known as the "complexity barrier" (Gros 2012: 183). The complexity barrier describes the situation wherein the effort of gaining scientific insight from a research field rises exponentially when approaching a certain threshold of complexity. If 'whole data' stands for the intellect of Laplace's daemon, then there may be data out there that are out of reach for any intellect other than such a daemon. The completeness of whole data is, therefore, not achievable, and consequently "big data and whole data are not the same" (Boyd & Crawford 2012: 669). Harford concludes with the following warning: "Found data contain systematic biases and it takes careful thought to spot and correct for those biases. 'n = all' is often a seductive illusion" (2014: 18). Big data will not lead to the theory of everything (Mainzer 2014).

2.1.4.4 *Taken out of Context, Big Data Lose Their Meaning*

Data are collected for a certain purpose, from a distinct vantage point, with distinct methods, and various tools (Kitchin 2014b). "However, in both its production and interpretation, all data – 'big' included – is always the result of contingent and contested social practices that afford and obfuscate specific understandings of the world" (Dalton, & Thatcher 2014). Several researchers conclude that raw data do not exist. In her edited book, Lisa Gitelman unmasks "'raw data' as an oxymoron" (Gitelman 2013). Bowker then explains that claim even further, judging that "'raw data' is a bad idea" (Bowker, 2005: 184). Data, and especially big data, are contextualized by the process of data generation (Kitchin 2014c) and, although the data set seems objective, distortion has already taken place in the process of collecting specific data. Information about the organization of data collection is essential (Lynch 2008), and, therefore, Maturana proposes that "anything said is said by an observer" (1970: 4). Applied to big data, the sentence could be transformed into: any data collected are collected by an observer. Von Foerster also introduced a variation to this assumption: "Anything said is said to an observer" (1979: 5) which could be adapted for big data as well: any data collected are collected for an observer. Data appear to always be firmly entangled with the entity that collects them. George et al. (2014: 322) categorize the sources of big data into the following five types:

- public data (collected by governments or governmental organizations)
- private data (collected by private-firms or individuals)
- data exhaust (passively collected and collected for a different purpose)
- community data (social media data)
- self-quantification data (collected by individuals to monitor or track themselves)

It seems obvious that there are reasons for collecting certain data, depending on the data collector and observer, but a certain tendency can be observed wherein big data "is [...] not used for the purpose for which it was collected" (Puschmann & Burgess 2014: 1699) and repurposing data has become common practice (Schneier 2015). The context of data collection is also often lost in transfer (Pasquale 2015), due to various types of invisible "access constraints and platform-side filtering" (Ruths & Pfeffer 2014: 1063). This fosters "the need for increased awareness of what is actually being analyzed" (Ruths & Pfeffer 2014: 1064). There is evidence of a self-selection bias, especially in the context of social media (Schoen et al. 2013). Taking Twitter social media data out of context, therefore, means assuming that Twitter generates a representative sample. This claim is flawed (Tumasjan et al. 2010) however, generating distorted results (Ruths & Pfeffer 2014).

In his reply, Seaver (2015) tackles the claim originally brought forward by Boyd and Crawford (2012) that big data can be completely taken out of context, and states that context is king, context is key, context is questioned, context is constructed, and context is contested. He explains that context derives current business, and emphasizes that context-aware systems do exist and actually work quite well. In his discussion, however, he predominantly focuses on the inadequacies of the term 'context' which is subject to constant debate (e.g. Dilley 1999, Johns 2006). In the context of big data, the term may lack precision. Boyd and Crawford (2012), and in particular Dalton and Thatcher (2014), Kitchin (2014b), and Pasquale (2015), highlight the problem that the transfer of data inevitably causes a loss of information. How were the data derived? Who collected the data? For what purpose were the data generated? These questions may sound like contextual information, but they also resemble the definition of metadata (or more precisely extended metadata).

> "To any data element, or to any of the component cells of a composite, can be associated, in a binary relationship, certain data elements which represent data 'about' the related element. We refer to such data as 'metadata' and call the relationship one of 'secondary association'" (Bagley 1968: 91).

For (extended) metadata, it will necessarily contain all the information about the observer and the methodological constraints that lead to a distortion of big data. These contextual aspects will be included in any data set in order to incorporate the influences brought by the data collection into other contexts. Losing, or getting rid of, this contextual metadata will change the data and will lead to a substantial increase in the number of statistical traps and obstacles.

2.1.4.5 Accessibility Does Not Make Them Ethical

Big data have led to an abundance of data available to anybody and obtainable without great effort (Fanning & Centers 2013). Many companies also act as data brokers (Otto et al. 2007) and sell a variety of data at low cost. People reveal more and more information about themselves (Enserink & Chin 2015), especially due to the recent trend of quantifying the self as a process of self-tracking aspects, such as running habits (Ruckenstein & Pantzar 2015). Aggravating the effect of the masses of data makes people slaves to their habits, and further facilitates their identification (Eagle & Pentland 2006). Some researchers (e.g. Tene & Polonetsky 2012) view big data as a contributor to the anonymization of the individual due to the sheer mass of data. The majority (e.g. Ohm 2010, Richards & King 2013, de Montjoye et al. 2015, Schneier 2015), however, would much rather discuss the "myth of anonymization" (Clemons 2013). Several companies have evolved, over time, to become proper "Datenkraken" (Bager 2006: 168), a German compound that translates to 'data kraken'. These companies acquire data from all available sources and use them in order to paint a granular picture of any person (Kucklick 2014). To make matters worse, every person leaves behind a trail of data. This is why privacy alone is no longer enough (Matzner 2014), because "privacy as we have known it is ending, and we're only beginning to fathom the consequences" (Enserink & Chin 2015: 491), especially as "any information that distinguishes one person from another can be used for re-identifying data" (Narayanan & Shmatikov 2010: 24). Schneier, an established security expert, takes a fairly drastic approach to demystifying this belief:

> "It's counterintuitive, but it takes less data to uniquely identify us than we think. [...] We can be uniquely identified by our relationships. It's quite obvious that you can be uniquely identified by your location data. With 24/7 location data from your cell phone, your name can be uncovered without too much trouble. You don't even need all that data; 95% of Americans can be identified *by name* from your four time/date/location points" (2015: 44).

Technological advancement will foster this development even further. One example is the location data generated by smartphones: due to the intimate relationship of humans with their smartphone (González, et al. 2008), people are now constantly accompanied by their phones. In consequence, data about the whereabouts of every smartphone user already exist and anonymizing it by removing personally identifiable information is absolutely insufficient. The amount of information that can be derived from meta data is even more shattering (Schneier 2015). Michael Hayden, former NSA and CIA director, was recently incensed enough to say that "we kill people based on metadata" (2014).

Big data have decreased and eradicated anonymity (Tucker 2013), and will be involved in the creation of a "goldfish bowl society" (Froomkin 2015: 130). Ethical use will be the most critical topic concerning big data, especially with regard to the numerous examples of the de-anonymization of large data sets. Two cases have been discussed in the literature in greater detail (e.g. Ohm 2010, Schneier 2015). AOL published search queries in 2006, and researchers were able to identify AOL users on the basis of this set of data (Barbaro & Zeller 2006). Netflix published their users' movie rankings, and researchers were able to de-anonymize people by comparing the data set with the Internet Movie Database (IMDb) (Narayanan & Shmatikov 2008).

The question is, therefore, no longer whether de-anonymizing big data sets is possible, but whether or not it is done. There are enough data available to conduct any analysis imaginable, but the ethical perspective has become prevalent. How do researchers (or big data analysts) act ethically (Boyd & Crawford 2012)? If big data are far from being anonymous, nobody can pledge privacy of data, but merely promise to use them in an ethical way. Mittelstadt and Floridi summarize this challenge in the context of biomedicine, but their statement holds true for any other discipline: "Data have been identified as particularly ethically challenging due to the sensitivity of health data and fiduciary nature of healthcare" (2015: 28) and, additionally, lead to "ethical responsibility development, deployment and maintenance of novel data sets and practices in biomedicine and beyond in the era of Big Data".

2.1.4.6 Limited Access to Big Data Creates New Digital Divides

If big data are really that ubiquitous and anonymization is not entirely possible, having access to data becomes a source of power. Andrejevic calls this phenomenon "the big data divide" (2014: 1673). Although everybody generates data, there are many with limited or no access to data at all. Manovich (2011) classifies these people into three stages of big data involvement: data creator, data collector and data analyzer. The difference in access, however, leads to "asymmetric sorting processes and different ways of thinking about how data relate to knowledge and its application" (Andrejevic 2014: 1676). Especially in the context of predictive analytics (Shmueli & Koppius 2011), people or algorithms with access to data can potentially influence those without access to such data (Palmas 2011).

> "When you are doing this kind of analytics, which is called 'big data', you are looking at hundreds of thousands to millions of people, and you are converging against the mean. I can't tell you what one shopper is going to do, but I can tell you with 90 percent accuracy what one shopper is going to do if he or she looks exactly like one million other shoppers" (Nolan 2012: 15).

Such processes lead to a form of social sorting (Lyon 2003) and, even worse, a sorting of people without access to data by people with access. Such presumed knowledge of the future can create self-fulfilling prophecies (Merton 1948), especially when people are nudged in a certain direction (Thaler & Sunstein 2008). Being able to execute this form of steering is a new and weighty source of power. Unfortunately, assessments of this kind, deliberate or accidental, are quickly written in stone:

> "For example, one data broker (ChoicePoint) incorrectly reported a criminal charge of 'intent to sell and manufacture methamphetamines' in Arkansas resident Catherine Taylor's file. The free-floating lie ensured rapid rejection of her job applications. She couldn't obtain credit to buy a dishwasher. Once notified of the error, ChoicePoint corrected it, but the other companies to whom ChoicePoint had sold Taylor's file did not necessarily follow suit. Some corrected their reports in a timely manner, but Taylor had to repeatedly nag many others, and ended up suing one" (Pasquale 2015: 33).

This shift in power means many people are growing more and more powerless when it comes to their data. They do not know what is collected, who is collecting, and for what purpose. There will be winners and losers (Richards & King 2013), there will be people who are empowered or disempowered (Mansell 2016). Current developments in big data increase the complexity as well as the opacity of the system (Burrell 2016) and will intensify this divide even further.

2.1.5 May Big Data Be with You

In summary, big data cannot be categorized into existing technological dimensions like data mining, algorithms and machine learning, or artificial intelligence. Big data are interconnected with those technologies and take a new form during this process. As artificial intelligence becomes smarter, more autonomous and opaque, big data are transformed in novel ways. Without big data and the abundance of data available, none of the current improvements in technology would be possible. IBM's artificial intelligence Watson, for example, is constantly learning from the internet (Madrigal 2013) and Google's artificial neural networks are capable of dreaming about their experiences within the internet (Mordvintsev et al. 2015). Those artificial intelligences, therefore, generate data on their own and could potentially learn from themselves. For that reason, disentangling big data from this technological cycle is as impossible as it is unnecessary. Big data have led to a quantum leap in those fields and, although those technologies merely mimic intelligence, big data make them smarter.

Big data are entangled in a complex way with data mining, algorithms and machine learning, and artificial intelligence. Big data enable those technologies to be better (Lohr 2012). On the other hand, big data are enabled by these technologies (O'Leary 2013). Big data contribute to a cycle of technology and can be depicted as in Figure 3.

Figure 3: Big Data's Technology Cycle

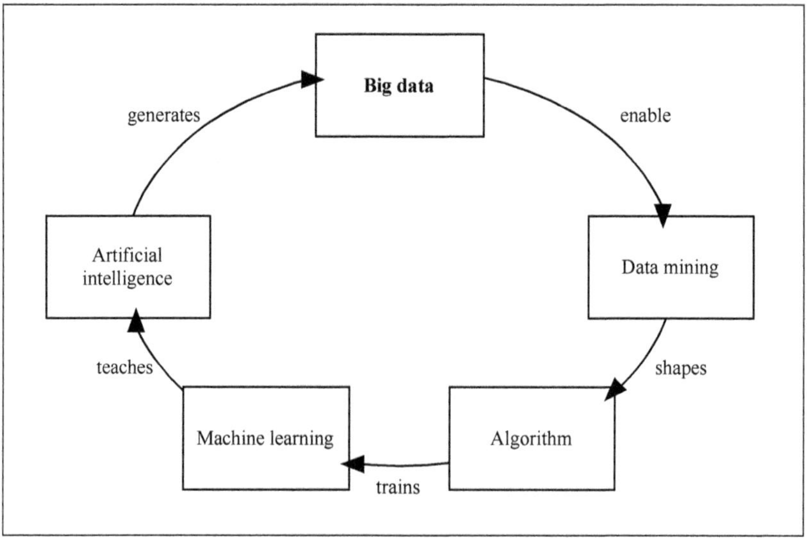

Consequently, big data are difficult to grasp, which renders defining big data a complicated task. The term 'big data' may not be sufficient for describing the phenomenon (Manovich 2011). "Big data is less about data that is big than it is about a capacity to search, aggregate, and cross-reference large data sets" (Boyd & Crawford 2012: 663). Big data, therefore, better represent a new grasp of data. Big data as a *new understanding of data* already have and will have in the future an unarguably significant impact on society. Digitization and technological progress are predominantly driven by data. Artificial intelligences work because of big data and mimic real intelligence quite well. Big data act as a lubricant for any advance in modern society. Big data are the basis of and a resource for modern communication (be it human to human, machine to machine, or human to machine) and transform the entire communicative process. Big data are "already occupying a huge place in the landscape of what technology is, what it might offer, and what it could be" (Bell 2015: 9). One could deduce that big data are everywhere (Cukier 2013). Big data have outgrown their lion's cage (Dietsch 2010). They float freely within the realm of both digital and analog worlds. The barriers between both worlds are also blurry and seem likely to merge in the future (Mayer-Schönberger & Cukier 2013, Kucklick 2014, Pasquale 2015). Big data will potentially contribute significantly to the theory of everything (Mainzer 2014). Metaphorically speaking, big data is a theoretical construct that surrounds everything, everywhere, and all the time. Big data are more than the *invisible hand* (Smith 1776) and sound similar to the description of 'the Force' in the Star Wars universe.

> "Well, the Force is what gives a Jedi his power. It's an energy field created by all living things. It surrounds us and penetrates us; it binds the galaxy together" (Obi-Wan Kenobi in Star Wars – A New Hope).

> "For my ally is the Force, and a powerful ally it is. Life creates it, makes it grow. Its energy surrounds us and binds us" (Yoda in Star Wars – The Empire Strikes Back).

In the context of big data, this sentence could be transformed into:

> Well, big data are what gives a human power. It's a data field created by all living things. It surrounds us and penetrates us; it binds the world together.

> For my allies are big data, and powerful allies they are. Life creates them, makes them grow. Their data surround us and bind us.

Although this metaphor may be a bit farfetched, it is crucial to state that big data are omnipresent, and many worldly things are no longer possible without *tapping into the realm of big data*. Big data are "here to stay" (Newell & Marabelli 2015: 10), "too big to ignore" (Simon 2013: xxi), and most importantly "here to grow" (Floridi 2012: 437). Contrary to the Star Wars analogy, however, where not all people can become Force-sensitive and use the Force, people can become big data-sensitive and use big data to their purpose. Big data in general are initially neutral, but as Gitelman (2013) asserts: "Raw data is an oxymoron". Data are shaped by a data-sensitive entity (human or machine), thus following the logic of Kranzberg (1986: 545): "Technology is neither good, nor bad; nor is it neutral" because "technology is a very human activity" (Kranzberg 1986: 557), however, big data "act as intermediaries in almost every aspect of our existence" (Seife 2015: 480). Due to their influence, big data cannot be seen as something purely technological, even though that perspective is already a difficult one. Beyond that, they are heavily entangled with today's society and, therefore, are understood as a socio-technological phenomenon. It is for that reason that "We must look beyond findings from previous studies of emerging technologies because new technologies often help create new socio-technological contexts" (Michael & Michael 2013: 26).

2.2 Big Data at the Socio-Technological Level

2.2.1 Technology and Society

Big data are highly entangled with society (Mayer-Schönberger & Cukier 2013). Society deals with big data on a daily basis and the influence of this datafication is growing exponentially. Technologies driven by big data are growing in significance. One example is how people find their way from A to B. GPS-powered smart-phone applications or dedicated devices have largely replaced printed maps.

> "It is easy to dismiss technology as a mere object, without giving much consideration to how it is woven into our everyday life. Technology provides a means for us, its users, to get things done. Without giving it much thought, we leave our homes every morning with our cellphones (often more than one), laptops, MP3 players (such as iPods), headphones, watches, and other gadgets. Only when our technology fails us we suddenly realize the depth of our dependence on that technology" (Quan-Haase 2016: 1).

For that reason, technology is more than a mere tool designed by society, instead it shapes and changes it. "We are no longer looking at just a 'technology' and its 'users' but the event of their relationships, of their reciprocal configuration" (Giddings 2006: 160). Technology challenges beliefs and structures (Heidegger 2011). Technology and society influence each other reciprocally (MacKenzie & Wajcman 1985). Both humans and technology are relational to each other as "equal" objects (Bryant, L. R. 2011). Technology are more and more interconnected and intertwined with human actors (Quan-Haase 2016). Although claiming the equality of humans and non-human objects as conscious actors may seem confusing, the hypothesis is legitimate; any object becomes part of the network due to its interaction within. Giving technology the freedom to work on its own (e.g. machine learning), makes it independent from the supervision of a human actor.

Technology on its own is no neutral thing. It is always connected to its network and, thus, connected to human actors (Kranzberg 1986). A plain piece of paper is nothing more than a potential tool. Writing on it, however, transforms it into something much more than a mere piece of paper. This new object evolves into data or maybe even information, and is integrated into a distinct network. Writer and paper are put into a relationship which, on top of that, is highly contextualized. Where was the paper written on? In what way? The mere use of an object (medium) already creates a message (McLuhan 1967), even without knowing what is written on the paper. The purpose of technology is constituted by its connection to reality (Whitehead 1929). On the other hand, it is a convention (Latour 2005) that connects various realities (Berger & Luckmann 1966), thus shaping the social habitus (Bourdieu 1977).

Technology in a broader sense can be categorized into "(1) material substance, (2) knowledge, (3) practice, (4) technique, and (5) society" (Quan-Haase 2016: 4). The first understanding disregards the connection between technology and society completely and views technology only as a passive tool under human control (Feist et al. 2010). The second definition compares technology to knowledge. Technological knowledge can be seen as a focus on the ability to create artifacts, objects that are created according to human intentions (Hershbach 1995), which re-presents the translation from technological ideas into designs, objects, and forms (Layton 1974). The third definition goes beyond the idea of creating artifacts and compares technology to a system, thus understanding it as a practice enclosed in human activities (Franklin 1999). Franklin uses this argument to reveal the negative potential of technology, especially in the context of becoming a normal part of society's routines and thereby changing people's behavior altogether. The fourth definition understands technology as a technique. 'Technique' is derived from the Greek word *technikos* and denotes human activity that possesses a certain goal and mechanism (Heidegger 1977). Technology is, therefore, not only a tool but rather a mechanism related to human activities (Ellul 1964). The final definition approaches technology with regards to society. Simpson (1995) sees technology as a change agent for society and some go even further: "It doesn't push things forward or transform the world, it becomes the world" (Baudrillard & Gane 1993: 44).

The evolution of definitions reveals that technology and society are indeed highly entangled. One could even propose a fusion of society and technology (Quan-Haase 2016). In terms of big data that claim may already be genuinely true, but even though they appear as two distinct actors within a network (Latour 2005), society influences big data and big data influence society. The relationship between big data and society is a deterministic one. Following this logic, sharpening the distinction between technology and society is essential to understanding the potential interferences between society and big data. Quan-Haase derived the following definition from a similar argument:

> "Technology is an assemblage of material objects, embodying and reflecting societal elements, such as knowledge, norms, and attitudes that have been shaped and structured to serve social, political, cultural, and existential purposes" (2016: 9).

On this basis, the interconnection and ubiquitous force of big data surrounding every human can be deconstructed and separated. This separation is essential to understanding both the technological deterministic and social deterministic viewpoints. According to the respective literature, this unilateral deterministic view falls into the category of hard determinism (Marx & Smith 1994). Both aspects are essential to deriving a better understanding of the socio-technologically intertwined relationship between big data and human actors.

2.2.2 Technological Determinism

According to the theory of technological determinism (e.g. Blauner 1964, Smith & Marx 1994), technology is the driving force behind social change. Any change in technology will cause society to adapt. Technological determinism identifies technology as the driving force behind social change in history (Kunz 2006) and is described as "the most important cause of change" (MacKenzie & Wajcman 1985: 4). This concept sees technology as independent and autonomous, guided by an internal logic (Quan-Haase 2016). The term 'technological determinism' was coined by Thomas Veblen (1921/2001) and built up on observations by Karl Marx: "The hand-mill gives you society with the feudal lord; the steam-mill, society with the industrial capitalist" (Marx 1971: 109). Although Marx criticized technological determinism and broadened it in order to include social productive forces (MacKenzie 1984), the theoretical debate substantiates the influence of technology: "The uses made of technology are largely determined by the structure of the technology itself that is that its functions follow from its form" (Postman 1992: 7). This determinism is strongly linked to the idea that "given the past, and the laws of nature, there is only one possible future" (van Inwagen 1983: 65). Moore's law (Adee 2008), for example, can be seen as self-directed and following its own form. The number of transistors is growing at an exponential rate, apparently without any influence from society. Technological progress in this field is determined by technology itself. To strengthen this potential of self-direction, Heilbroner (1967) points out the phenomenon of simultaneous discoveries (Merton 1961, Bikard 2012) and, consequently, supposes a general predictability of technology (Bellow 1959, Martin 2010). Heilbroner reports the absence of sudden technological leaps. This argument, however, may lose its validity in modern times (Sood & Tellis 2005). Technological life, consequently, dominates social, political, and economic life (Ellul 1964), as well as technological norms of practices, under the aegis of rationalization (Habermas 1970). This perspective, though frequently described as bold (e.g. Bimber 1994), describes the idea that society adapts to technological change (Miller 1984). Winner defines technological determinism as a form of technological somnambulism (2004). He asserts that the behavior of society compares to sleepwalking when it comes to technology. Technology creates new worlds and restructures everyday lives. Interestingly, Winner (2004) highlights that people tend to be blinded by the usefulness of technology and, therefore, do not realize the societal transformations brought upon them by technology. Technology acts as a change agent for changes that often go unrecognized by society.

The general idea behind technological determinism seems insufficient when trying to grasp the interaction between technology and society (e.g. MacKenzie & Wajcman 1985, Degele 2002), but is still popular as it conveys an understanding of the force of technology in today's world, especially due to the assumption that technology will eventually solve today's problems, and deliver a technological fix (Drengson 1984) capable of clearing the way to a technological utopia (Segal 1984). Interestingly, the idea of a technological fix re-emerged in the discussion about big data. Morozov criticizes the belief that every complex problem is solvable "if only the right algorithms are in place" (2013: 5), for which he coins the term 'technological solutionism'. His argument underlines the potential of big data to be technologically deterministic. The effect seems increasingly prevalent because big data can be opaque, obscure, and overwhelmingly complicated. Big data as a type of technology appears about "to slip from human control" (Heidegger 1977: 289).

Big data have an inherent ability to enforce social change, which is clear in the current applications of big data. Based on algorithms, big data generate a certain and distinct form of reality. One aspect of changing reality is the filter bubble (Pariser 2011) that shields society from certain data. Chomsky (2013) clarifies: "We can find lots of data; the problem is understanding it. And a lot of data around us go through a filter so it doesn't reach us." Due to the abundance of data, people tend to over-estimate the effect of data drastically, and are more inclined to rank results as objective and, consequently, truthful. Big data have a tendency to mirror reality (Bolin & Schwarz 2015). This is why big data construct social reality, hiding behind a veil of objectivity and granularity, but are in fact adhesively subjective and imprecise. This creates a kludge that is comparable to an assemblage (Kitchin & Lauriault 2015) or mosaic (Sprague 2015) of reality. Society, however, recognizes this mosaic as reality which makes big data representative of a type of social constructivism or, to be more precise, a form of "data constructivism" (Scholz 2015a: 10).

On the basis of data constructivism, social cognition is guided by big data and the underlying algorithms. Algorithms, however, always include an ideology of sorts (Mager 2012). Such an ideology is not bound exclusively to subjective measures, but similarly connected to statistical measures. Pentland (2014) argues that society follows certain rules and abides by statistical regularities. Such regularities can be seen by means of big data and, therefore, used to optimize society (Mayer-Schönberger & Cukier 2013). Carr (2014), however, criticizes this approach as follows: "Pentland's idea of a 'data-driven society' is problematic. It would encourage us to optimize the status quo rather than challenge it" (2014), as it could potentially result in a deadlock (Rätsch 2015). Big data will transform into civic "thermostats" (McLuhan 1967: 68) while declaring the ideal temperature (Carr 2014). People may become "slaves to big data" (Hildebrandt 2013: 1).

Social engineering of this type strongly influences people's behavior, not because the data are extremely precise, but because it shows a socially accepted path. People are nudged (Thaler & Sunstein 2008) into a distinct direction which renders them more willing to follow that path. Anticipatory obedience (Lepping 2011) turns that prediction into a self-fulfilling prophecy (Merton 1948) or even self-preventing prophecy (Brin 2012). Such a prediction merely needs to appear probable for it to actually happen. Data constructivism has substantial power over people and will lead to a form of social control (Scholz 2015a). Any data-driven society is determined purely by the use of available data, which may be insufficient. While statistical analyses may be rigorous, they may lead to completely different results following either the Gaussian or the Paretian statistical principle (Bolin & Schwarz 2015). Even though the results are objective and statistically correct, they are only a chimera of reality (Brenner 2013).

Assuming that big data shape society leads to a data constructivism of reality. Even though this may sound promising, a deficiency of big data will not fix all societal problems but will lead to a specific type of social engineering. People more and more commonly act according to rules that are disguised as socially acceptable. Frischmann (2014) even proposed a Reverse Turing Test, which does not aim at measuring the ability of a machine to act like a human being, but rather a human's ability to resemble a machine. The test is intended to reveal whether humans are, indeed, controlled by nudges from technology such as computers, smartphones or wearables (Yeung 2016).

All things considered, big data provides society with as much information as possible, consequently overwhelming it (similar to Huxley 1932). It may be so overwhelmed, in fact, that society as a whole is nudged towards an entity, such as an algorithm that selects "relevant" information (similar to Orwell 1949) for society. Such algorithms conceal the objective truth from society on the basis of big data and, perhaps not even purposefully, result in the data constructivism of reality.

2.2.3 Social Determinism

The opposite relationship is social determinism (e.g. Pannabecker 1991, Green 2001). Supporters of this concept believe that society creates technology in order to fulfill a certain need in society (Quan-Haase 2016). This theory identifies people as central drivers of change. As a result of societal change and the corresponding societal needs, people develop new technologies which then lead to technological progress. Contrary to technological determinism, technology is not autonomous and self-directed (Winner 1993). Society or social groups attribute meaning to technology and its use or impact. For that reason, technology is influenced by a variety of social factors such as history, economics, and ideology (Giddings 2006). Technology is a social construct that receives its meaning and relevance from society (Winner 1993). Following that idea, technology is being developed to saturate society's needs, thus overcoming human limitations.

The social deterministic view has spiked several advances within the discourse, and has contributed to the emergence of the academic field "science and technology studies" (influenced by the seminal work of Kuhn (1962)). The academic field rapidly rejected both technological determinism and social determinism (Bijker et al. 1999), and realigned its focus on the mutual shaping (Boczkowski 2004) of society and technology, however, in the beginning of the field, some researchers called this social deterministic influence the social construction of technology (e.g. Bijker et al. 1987). They also proposed a superiority of society over technology in that it is society that shapes technology. The use of technology is also influenced by social context (Klein & Kleinmann 2002). Pinch and Bijker (1984) define this concept on the basis of four key terms: (1) relevant social group, (2) interpretative flexibility, (3) closure and stabilization, (4) wider context. The authors argue that technology can only gain meaning and consequently survive when receiving societal support (Pinch 2009). A lack of support or interest from society prevents the development of a certain technology, or as the authors state: "a problem is only defined as such, when there is a social group for which it constitutes a 'problem'" (Pinch & Bijker 1984: 414). On this basis, interpretative flexibility suggests that technology is not neutral, as its meaning can vary according to social context. This can be exemplified by the discovery of microwave ovens. They were accidently discovered in a radar station when, by coincidence, a melted candy bar was found in close proximity to the source of radar waves (Andriani & Cohen 2013). Microwaves as a specific technology were moved from one socio-cultural context into a completely different one by a certain social group. While the technology (microwaves) remains the same, its meaning for society obviously differs drastically depending on whether the waves are used for tracking planes or cooking food. This change in usage can be described as cultural transduction (Uribe-Jongbloed & Espinosa-Medina 2014). Andriani and Cohen (2013) also established the concept of closure and stabilization. Closure is reached once a social group comes to an agreement about the purpose of a certain technology, and stabilization is achieved when the technology seems ready for the market. Finally, there is the wider context. Even if a certain social group supports and demands a new technology, it may still be rejected. It seems obvious for example that the future belongs to the electric car, however, many people still currently reject it (Pierre et al. 2011). Various social groups support the electric car, and it tackles a specific societal need, but in the wider context a majority choose a different technology (Winner 2003).

Following the idea of social construction, Winner (1993) links research into the social construction of technology with social constructivism. Lawson (2004) goes further by arguing that any social-deterministic view is rooted in social constructivism. Following this argument, social determinism can be linked to big data as a technology, and big data understood as being shaped by a form of social constructivism. Social constructivism, therefore, serves as the counter-argument to data constructivism. One essential aspect of the following discourse is that there are social groups that shape technology, and as Winner (1993) reports, those groups will take over elitist roles and obtain great power. Floridi (2012) reasons that the game of using big data will be won by those with the ability to use big data and, as he quotes from Plato, "those who 'know how to ask and answer questions' (Plato, Cratylos, 390c)" (Floridi 2012: 437). Floridi raises a crucial concern that big data will lead to a big data divide (Andrejevic 2014) and that the people with the power, as well as the ability to use big data, will shape the social reality of those without access to, and knowledge of, big data. Boyd and Crawford (2012) separate society into big data rich and big data poor. The big data rich will have the power to use social constructivism.

The most defining work regarding social constructivism was written by Berger and Luckmann (1966) who claimed that society can be seen as both an objective and subjective reality. In their sense, however, objective did not carry the same connotation as the commonly used term 'objectivity' as the authors clarify: "It is important to keep in mind that the objectivity of the institutional world, however massive it may appear to the individual, is a humanly produced, constructed objectivity" (Berger & Luckmann 1966: 60) and, to be more concise that "society is a human product" (1966: 61). As far as subjective reality is concerned, this mainly focuses on the reality constructed by individuals on their own in developing their subjective reality through socialization and interaction with nature. Berger and Luckmann elaborate on the idea as follows: "Man is biologically predestined to construct and to inhabit a world with others. [...] man produces reality and thereby produces himself" (Berger & Luckmann 1966: 183).

The authors claim that objective and subjective reality co-exist and mutually shape social reality, however, the process of social constructivism follows a certain logic. Social reality is created through a certain form of externalization. A particular social group comes to a preliminary consensus, thus creating a set of norms which in repetition becomes habitualization. If this constructed reality is reproduced by others, it becomes institutionalized, will be aligned with new institutions, and form a shared language. This stage is followed by a phase of objectification during which the construct of reality receives legitimation. This institutionalized and legitimated reality will now be passed down from generation to generation and consequently internalized by society. The result of this process is human-constructed objectivity.

When considering only a small social group, the assumption of such processes of social interaction constructing reality appears plausible. New technologies like big data, however, can be used alongside the constitution of social reality. Big data can help repeat certain social realities, dominate the language in use, reveal the need for certain institutions, support the process of legitimation and transfer such a reality to other individuals and generations. Big data are, therefore, a powerful tool with which to internalize a certain social reality. Such effects are intensified by changes in society's communicative behavior. Berger and Luckmann have already envisioned a future that seems similar to today's digitized communication:

> "The social reality of everyday life is thus apprehended in a continuum of typifications, which are progressively anonymous as they are removed from the 'here and now' of the face-to-face situation. At one pole of the continuum are those others with whom I frequently and intensively interact in face-to-face situations – my 'inner circle', as it were. At the other pole are highly anonymous abs actions, which by their very nature can never be available in face-to-face interaction. Social structure is the sum total of these typifications and of the recurrent patterns of interaction established by means of them. As such, structure is an essential element of the reality of everyday life" (Berger & Luckmann 1966: 33).

In the future, structure will be an essential parameter for social constructivism and a process of structuring is already ongoing. As previously explained, big data divide society into two groups, with the elitist group shaping the use of big data. One group will always be using new means for their own purposes. Such behavior is commonly referred to as Campbell's law:

> "The more any quantitative social indicator is used for social decision-making, the more subject it will be to corruption pressures and the more apt it will be to distort and corrupt the social processes it is intended to monitor" (Campbell 1979: 85).

While the question of whether big data would knowingly be used for harmful purposes may be debatable, there is a temptation to use big data to nudge people's behavior in a certain direction (Schroeder 2014). One popular example is the potential of the media to steer, manipulate, and control social reality (e.g. Chomsky 2002). Despite great controversy, differences in coverage inevitably change the narrative of that coverage as observed, for example, in the context of climate debate (Feldman et al. 2012). Selecting certain information and omitting other information can be used to change social realities. This effect becomes even more pronounced when not all available information can be accessed. Society is "drowning" (Puschmann & Burgess 2014: 1699) in data and seeking any available reduction in complexity. Such a reduction in complexity is controlled by different social groups and, therefore, entangled with certain agendas and ideologies (Mager 2012).

Information is selected by certain social entities in order to nudge people into certain behavior (similar to Orwell 1949). People unconsciously commit themselves to a social entity, as they are overwhelmed, and scared away from any other social entity (similar to Huxley 1932). Such social groups shape their own realities, and have the ability to spread and impose their reality on other people, resulting in a social constructivism of reality.

2.2.4 Socio-Technological Concurrence

Speculating about the effects of technology on society and those of society on technology eventually reaches its limits. This is why the interdisciplinary field of science and technology studies has developed a stance of general rejection towards technological determinism and social determinism and instead focuses on the mutual shaping of technology and society. Technology cannot be seen separately from society, much as society does not evolve independently from technology. Both are integral parts of a holistic socio-technical system (Bijker et al. 1999). A first approach to describing the relationship between technology and society is technological momentum (Hughes 1969). Hughes uses technological and social determinism but connects both models from the perspective of time. He claims that "a technological system can be both a cause and an effect; it can shape or be shaped by society" (Hughes 1994: 112). Over time, a technological system will move between the extremes that are technological determinism and social determinism. Hughes reasons that a technology is shaped by society at first, but over time evolves into a technology shaping society. It gathers *momentum*.

Elaborating on previous research, Callon, Latour, and Law developed Actor Network Theory (ANT) in the 1970s (Murdoch 1997). Building upon the concept of mutual shaping, ANT understands technology as dynamic process (Latour 1987). It therefore influences the social network, which renders it a part of the network. The theory attributes to technology the ability to act in a certain way, however, and influence the network. In this way, it varies from the perspective of social constructivism (Giddings 2006). As Latour highlights by himself: "[ANT] entirely bypasses the question of 'social construction' and the realist/relativist debate" (Latour 1999: 22). ANT is, thus, incompatible with theories such as the structuration theory by Giddens (1984) and, therefore, does not follow the duality of technology within organizations (Orlikowski 1992). The relationship is described as follows:

> "If human beings form a social network it is not because they interact with other human beings. It is because they interact with human beings and endless other materials too [...] Machines, architectures, clothes, texts – all contribute to the patterning of the social" (Law 1992: 382).

ANT links human beings and non-human beings and goes even further in the sense that it ascribes to those non-human beings (any kind of technology) the role of an actor (Latour 2005). Both share a form of social assemblage and "enter a stable definition of society" (Latour 1991: 129). MacKenzie and Wajcman (1999), interpreting the relationship between society and technology as described by ANT, conclude that they are made out of the same material, thus linking human and non-human actors in the same way. Latour underlines this claim in claiming to "see only actors – some human, some non-human, some skilled, some unskilled – that exchange their properties" (Latour 1992: 236). Other researchers agree, such as Bryant (2011a), for example, who understands both humans and technology as part of a relational network and, within the network, equal objects. Giddings (2006) dissolves the difference between technology and users, which represents a shift from traditional viewpoints. Latour (2005), however, uses the example of an airplane. The airplane is not only controlled by a pilot, but by an uncountable number of different actors that, within their relational network, keep the airplane flying. The focus shifts from the question of why relationships between actors exist (human and non-human beings), to the question of how these relationships work (Quan-Haase 2016).

When ANT is applied to big data and society, it fulfills the need for mutual shaping. The idea that big data and humans within the societal network are viewed from a relational perspective (Giddings 2006) aligns with the analogy of big data being a force surrounding everybody. Connecting big data to society in such a way mimics the influence of big data on social reality. This thought does not entail that one dominates the other or vice versa, so much as both actors enriching each other, which leads to a form of socio-technological concurrence of big data and society, mutually contributing to the construction of reality.

Proposing a certain concurrence, however, means that technology and society are changing, growing, and evolving together, but not at the same speed. Big data in particular are growing at a pace so mind-boggling that people compare big data to an avalanche (Miller 2010). The videogame "Mass Effect 2" (issued by Bioware in 2012) includes a quote by Mordin Solus (he is an alien which accounts for his unusual English) that targets the problems of such an imbalance:

> "Disrupts socio-technological balance. All scientific advancement due to intelligence overcoming, compensating, for limitations. Can't carry a load, so invent wheel. Can't catch food, so invent spear. Limitations. No limitations, no advancement. No advancement, culture stagnates. Works other way too. Advancement before culture is ready. Disastrous."

Solus confirms the concurrence of technological and societal progress and underlines the importance of a certain balance between both aspects. In a nutshell, big data can be seen as an actor within the social network, shaping society as much as society shapes big data. There is, however, a need to master this new technology (Heidegger 1977). Merely making big data an actor within the societal network would not suffice, as they would be too raw and too vague. In a next step, therefore, big data need to be explained from within an organization and by means of existing organizational theory. This is especially reasonable considering big data's current reputation as merely being a form of "unclear technology" (Cohen et al. 1972) within organizations.

2.3 Big Data at the Organizational Level

2.3.1 Epistemological Framing

The previous chapters reveal the difficulty of grasping big data as a theoretical construct. Big data are not clearly outlined as a type of technology separated from society, but need to be treated as technology that interacts intensively with society. In fact, the meaning and influence of big data evolve through this delicate interaction. There is no deterministic direction but a concurrence between both actors. There is also a certain friction between big data and society, which is why their relationship mimics duality. In this case, duality denotes the instance of both functions seeming contradictory, while in fact being complementary (Evans & Doz 1992). Janssens and Steyaert define this as follows:

> "Duality has the most general meaning of the three concepts: paradoxes and dilemmas can be seen as dualities, but not all dualities can be seen as paradoxical or simultaneously contradictory, or involving an either-or situation or an impossible choice" (Janssens & Steyaert 1999: 122–123).

The duality between big data and the members of an organization are even more apparent. Despite this, the task of grasping big data as part of an organization is more complicated. Big data are not something that can be 'gripped' and, therefore, are not always restricted to one organization alone. But there are what is referred to as 'organizational big data', big data that are uniquely contextualized towards one particular organization. When trying to explain big data within an organization, on the basis of organizational theory, one criterion to pay attention to is the closeness and openness of a system. The following chapters outline a series of theories located somewhere between predominately closed systems and completely open systems, and searching for theoretical explanations of the behavior of big data within an organization.

Several preceding comments on big data and organizational theory are necessary, especially concerning three assumptions that recur in the discourse of organizational theory, and that will be distorted by the implementation of big data within organizations: the assumption of bounded rationality, the modernist and postmodernist view of organization, and the discourse about the iron cage.

The term *bounded rationality* was coined by Simon (1959) and describes the fact that people make decisions on the basis of their limited information. Based on knowledge gained from incomplete information, people use heuristics to decide things and, therefore, sometimes act irrationally. People decide within certain constraints, which can be without apparent contextual, procedural reason or result from various other influences (March 1978). People lack the ability to make perfect rational choices (Pescosolido 1992) and are only capable of choosing the option that seems best to them. Allison (1969) vividly illustrates this in his analysis of the decision process during the Cuban Missile Crisis. People are bound by their incomplete rationality and especially by the limitations of the information available to them. Big data increases the amount of available information and, therefore, apparently contributes to more rational decision making.

Such a conclusion, though on a smaller scale, was discussed by March (1978) who highlights the risk of self-evident empirical truths. People tend to fall for the "illusion" of rationality that accompanies big data, but in order to form a basis for rational choice, big data need to be a precise image of reality. Its range, however, although far beyond that of an individual, is limited to only a small portion of reality. In the words of Wittgenstein: "The limits of my data [originally: language] mean the limits of my world" (1922: 74), but since big data are incomplete, any decision on the basis of big data will always rest upon bounded rationality. No organization on its own will have access to all big data, which is why any organization is regulated by bounded rationality. Decision making within an organization is, therefore, not only subjugated by the individual bounded rationality of people, but also by the bounded rationality that derives from accessible and available big data. Such *bilateral bounded rationality* establishes the constraints of organizational bounded rationality.

The next assumption regarding organizational theory is that theories can be categorized into a *modernist* and *postmodernist approach* and, above all, describe organizations differently (Cooper & Burrell 1988). Boisot and McKelvey (2010) proposed the provocative idea that modernist and postmodernist perspectives can be bridged. In times of big data, it may be a compelling argument that there is a need to bridge these perspectives. The authors connect modernism to positivism and, thus, link empirical observation with objectivity. This is only possible through repetition and the replicability of events. This is often interconnected with the Cartesian view (e.g. Miller 2008), according to which any cause brings about an explainable effect as well as some form of stable environment. Although such a reductionist view has been vigorously challenged (e.g. Alvesson & Kärreman 2011), Gaussian statistics focusing on normal distribution are still widely applied in the (social) sciences (e.g. Greene 2003). The goal of modernism is to produce robust and objective knowledge, but the postmodernist critique is that the stories derived from the acquisition of knowledge (Calás & Smircich 1999) are socially constructed stories. The social world consists of a form of radical subjectivity (Foucault 1977) and is influenced by power (Townley 1993), the scope of interpretation (Latour 1988), or regional and cultural contexts (Soja 1999). On the basis of this idea Boisot and McKelvey (2010) state that postmodernist distrust laws that are derived from normal distributions.

Behind every result there is a story or narrative that leads to "infinite conversations" (Wyss-Flamm & Zandee 2001: 297). As opposed to a reductionist view, postmodernists try to engage in social complexity (Cilliers 1998). Consequently, there is no way for any researcher to find one objective truth, at least not in the social sciences. The authors see a way "to integrate the ordered world of modernists and the more 'chaotic' world of postmodernists" (Boisot & McKelvey 2010: 416), especially as both worlds describe an atomistic and a connectionist ontology that may be scalable.

In the context of big data, scalability is the clue. Big data can be granular, detailed, and precise, but also be general and universal. Big data can record the interactions of organizations on a global scale, as well as the interactions of individual employees on a local scale. In the context of an organization, particularly, a story or narrative can be monitored over time by means of big data (Kim et al. 2013), which may result in a unique organizational signature (Stein et al. 2016). Such use of big data, however, depends on the combination of both the modernist and postmodernist perspectives. On the one hand, the modernist view delivers an over-generalization in the sense that all organizations seem similar. On the other hand, the postmodernist view would lead to a conception of organizations, it was so contextualized that a comparison would be altogether impossible. Big data are not merely a tool to bridge both views, but also provide organizations with a form of *scalability* between both views.

The final assumption to be discussed before analyzing the theories is the existence of the *iron cage*. In 1952, the first English translation of Weber's book "The Protestant ethic and the spirit of capitalism" used the term iron cage. However, in the German original Weber was talking about *stahlhartes Gehäuse* which more precisely translates into "shell hard as steel" (Baehr 2001: 153). Baehr traces back this fundamental change in meaning to the free interpretation by the translator Parson. The author elaborates, furthermore, that cage means being trapped in something, but shell describes a "living space both for the individual who must carry it around" (Baehr 2001: 163). In a certain way, the actual meaning of Weber is comparable to an augmentation of the organization and as something the organization can carry around. Such a shell could be beneficial or harmful, but such a metaphor would "appear anticlimactic" (Baehr 2001: 164). Nevertheless, the term iron cage became popular in social science (Baehr 2001) and, therefore, is used commonly for describing the situation in which organizations and their members are caged within a bureaucratic rationalization. Organizations are shackled by a precisely organized and mechanically tuned bureaucracy and, due to its apparent superiority, other organizations tend to adapt to such a rationale. DiMaggio and Powell (1983) agree that organizations have a tendency to homogenize their structures, however, they do not ascribe the homogenization to bureaucracy but to structuration (Giddens 1979). The basis of this is the question of why there are so many similar organizations. DiMaggio and Powell (1983) claim that this homogenization can be described as isomorphism. They follow the description of Hawley (1968): "Isomorphism is a constraining process that forces one unit in a population to resemble other units that face the same set of environmental conditions" (DiMaggio & Powell 1983: 149). Bureaucracy is only one reason for isomorphic change. Other strong influences are the distribution of power and the social legitimacy of an organization in comparison to other organizations. DiMaggio and Powell (1983) list the following mechanisms as driving forces for institutional isomorphic change. (1) *Coercive isomorphism* refers to institutions that pressure organizations in a certain direction. Although the introduction of new laws is the most prominent example, informal pressures are likewise possible. (2) *Mimetic pressure* is the belief that an organization becomes safer or more stable by mimicking another organization. (3) *Normative isomorphism* denominates structural change due to the professional background of an organization's members. Similar educational backgrounds encourage isomorphism. All three factors have an influence on the isomorphic tendency of one organization.

Big data can contribute to the isomorphic tendency of an organization through any of these three factors. External pressure is reinforced by supportive data, which leads to more coercive isomorphism. Big data reveal the fittest companies and provide enough information to mimic another organization completely, which, therefore, increases mimetic pressure. Big data also allow everybody to gain knowledge. Said knowledge, however, underlies a certain homogenization, as well as a form of Westernization (Wilson et al. 2006) which boosts normative isomorphism. Big data open up new sources of information to an organization. Assuming that big data contribute to anti-isomorphic tendencies, however, would be a false conclusion. Overall, big data increase and reinforce isomorphic tendencies, metaphorically securing the iron cage. Big data, being incomplete, lead to data constructivism and the creation of a certain reality. This increases the isomorphic tendency in a certain direction in accordance with this one data-constructed reality. The use of big data also gives any institution power and legitimation due to an apparent trust in numbers (Porter 1996). In addition to being reinforced, the iron cage becomes transparent. In their original article, DiMaggio and Powell (1983) give directions for those institutions that influence the isomorphism within an organization in a similar way to Bentham's panopticon (1843), in which the inmates of a prison are watched by a number of watchmen on a tower at all times. Organizations know that there is isomorphic pressure and, thus, adapt to normative expectations. In times of big data, however, organizations find themselves facing a post-panoptical scenario (Baumann 2000). In this case, organizations no longer know who is pressuring them and where the isomorphic tendency is directed, but the need to adapt to other organizations in the sense of homogenization remains obvious. When big data single out a certain type of organization as beneficial, organizations will change to match this structure, unaware of who decided it, and how this institution came to that conclusion.

In summary, all three assumptions reveal that big data are part of any organization and will play an integral role in understanding organizations. However, big data construct a new layer of reality within organizations. Big data cannot be seen as a mere source of information and, therefore, an external and objective factor, but much rather as an internal and subjective factor. Big data will contribute to certain solutions and intensify other problems. As the complexity of big data and their implementation increases, big data do not stop at the border of an organization as they are becoming increasingly heterogeneous.

2.3.2 Organizations as Open Systems

Generally speaking, organizations are never completely closed systems which marks a major difference to other fields of research. Organizations always interact with their environment to a certain extent. Von Bertalanffy (1968) was, in the 1940s, one of the first to describe the difference between closed and open systems. While it is necessary to define closed systems within the realm of physics, all other systems that are organized in any form will differ because they interact with other organizations from the outside.

> "However, we find systems which by their very nature and definition are not closed systems. Every living organism is essentially an open system. It maintains itself in a continuous inflow and outflow, a building up and breaking down of components, never being, so long as it is alive in a state of chemical and thermodynamic equilibrium but maintained in a so-called steady state which is distinct from the latter" (von Bertalanffy 1968: 39).

In the social sciences particularly, organizations are never closed systems, but systems that are living or social (Luhmann 2011). Luhmann (2011) also describes a different form of closed systems: a system can be operationally closed, which means that there is an outer side to an organization that faces the environment, as well as an inner side of an organization that does not interact with the environment and conducts tasks and operations completely independently of it. Contrary to a completely closed system, only the operational tasks are separated from the environment. Normally, such operational tasks are precisely described and there is no need for external interaction. For example, the production of a sheet of paper takes place within an organization and without interaction with the environment and, therefore, is operationally closed, though everything else is done in interaction with the environment. Separating the operational perspective from the general system is reducing complexity and enables the researcher to focus on the observation of organizations.

Table 6: Overview over the Theories on Open Systems

Operationally Closed System		**Open System**
Cybernetics	Systems Theory	Population Ecology Theory
Complex Systems Theory		

In order to integrate big data into the network of organizational theories, I will use the structure shown in Table 6. All theories mentioned there have increasingly opened towards the environment. All three selected theories contribute to a dynamic perspective on organizations and are linked with the complex systems theory. One final preliminary remark about the differentiation between cybernetics and systems theory: both theories are relatively similar and, therefore, the terms are often used synonymously (von Bertalanffy 1972). As a matter of fact, it is sometimes difficult to attribute a certain concept to one distinct field of theory and the correctness of the following selection of concepts is subject to debate. In the course of this thesis, however, both theories will be differentiated according to one specific aspect: cybernetics considers a system more from a predominantly technical or mechanical perspective (Ashby 1956), and systems theory rather from a social or organic one (Luhmann 2011). I assume that cybernetics will contribute more towards understanding the effect of big data on a social system, while systems theory will likely contribute to understanding the effect of a social system on big data.

2.3.2.1 Big Data in Cybernetics

The term cybernetics goes back to Wiener who, in the title of his corresponding book, defines them as "control and communication in animal and machine" (Wiener 1948). The term is derived from the Greek word *kybernētēs* and means steersman or pilot. Ashby adds to this by explicitly characterizing cybernetics as "the art of steermanship" (1956: 1). He also specifies it as "theory of machines" (1956: 1), but moves away from merely describing the machine in favor of trying to understand its behavior. Rooted in such mechanical thinking, cybernetics has influenced the computer sciences (Umpleby & Dent 1999), robotics (Arkin 1990), simulations (Forrester 1994), and the internet (Licklider 1960). The theory was also expanded to social systems and had a strong impact on the understanding of organizations (Morgan 1982). A prominent example of the use of cybernetics in a social system was the steering of Chile in the 1970s by a cybernetical system called CyberSyn, as envisioned by Stafford Beer (Medina 2006).

Cybernetics can be categorized into first order cybernetics and second order cybernetics (von Foerster 1979). The main difference is the role of the observer in the respective systems; in the first order, the focus is on the observed system. The second order, however, focuses on the actions of the observer. Umpleby (1990) summarized several definitions as depicted in Table 7, and expands the general definition by stating that first order cybernetics involves focusing on the model of a controlled system, and second order cybernetics makes the modeler central and treats the system as something autonomous. He contributes his own definitions which highlight the differences in interaction and the differences in the use of theory. Cybernetics shifted from a realistic or positivist view towards a more constructivist perspective (von Glasersfeld 1979).

Table 7: Definitions of First and Second Order Cybernetics

Author	First Order Cybernetics	Second Order Cybernetics
von Foerster	The cybernetics of observed systems	The cybernetics of observing systems
Pask	The purpose of the model	The purpose of the modeler
Varela	Controlled systems	Autonomous systems
Umpleby	Interaction among the variables in a system	Interaction between observer and observed
Umpleby	Theories of a social system	Theories of the interaction between ideas and society

(Umpleby 1990: 113)

One key concept within first order cybernetics is the law of requisite variety (Ashby 1956), which is often reduced to the quote that "only variety can destroy variety" (1956: 207). In more detail, Ashby explains that a fixed amount of variety is imposed by an external player D (disturbance) and that there is a variety of responses to come from a player R (response). He explains that "only variety in R's moves can force down the variety in outcomes" (1956: 206). Any R is capable of regulating the variety of outcomes due to the external input of variety by D. However, the "capacity as a regulator cannot exceed R's capacity as a channel of communication" (1956: 211). Ashby also states that there is the "hard external world, or those internal matters that the would-be regulator has to take for granted" (1956: 209), which he calls T (table). Thus, T is influenced by the variety of D and regulated by R. R is of utmost importance in regulating D and T, in order to influence the outcome.

Transferring this concept to organizations, there is a variety of external input as well as a variety of responses from organizations, that will lead to a variety of outcomes. Boisot and McKelvey (2010) call the spectrum of variety 'the Ashby space'. They propose the idea that an organization deals with a variety of stimuli and has a variety of responses. Both varieties can be low and high in this model. As defined by Ashby, however, a high variety of stimuli will lead to a high variety of responses. The regulation of variety is imposed by some form of ordering principle that tackles T – the authors use algorithmic compression as an example (Boisot & McKelvey 2010) – which makes it possible to categorize the Ashby space into an ordered regime (low variety of stimuli and low variety of responses), complex regime (medium variety of stimuli and medium variety of responses), and a chaotic regime (high variety of stimuli and high variety of responses).

Applying this concept to big data in an organization helps in understanding the general effect of big data within an organization. Big data can be seen as an external force that is taken for granted by the organization as well as an external force that disturbs the organization. Big data contribute massively to the variety of stimuli and, if unfiltered, will lead to a massive increase in the variety of response and will exceed the communication capacity of any organization. The organization will require the capability to "block" (Ashby 1956: 212) harmful big data and let the beneficial big data in. The organization will respond towards big data and incorporate useful and relevant information. Consequently, there are big data that may be beneficial but also that may be harmful for the organization. Ashby, as well as Boisot and McKelvey (2010), highlight that such regulation is the task of an organization, especially where the response of regulation influences the chances of an organization's survival (Ashby 1956). Big data will be regulated and ordered in some form in order to destroy variety and lower the variety of outcomes. The law of requisite variety claims that big data will be regulated by the organization itself and not by any external source, thus enforcing the idea that any organization *will deal with their own big data on their own.*

Another popular concept in first order cybernetics is homeostasis (Wiener 1948, Ashby 1952, Boulding 1956), based on the homeostasis concept as introduced by Cannon (1926). He defines homeostasis as follows:

> "The highly developed living being is an open system having many relations to its surroundings [...]. Changes in the surroundings excite reactions in this system, or affect it directly, so that internal disturbances of the system are produced. Such disturbances are normally kept within narrow limits, because automatic adjustments within the system are brought into action, and thereby wide oscillations are prevented and the internal conditions are held fairly constant. The term "equilibrium" might be used to designate these constant conditions. [...] The coordinated physiological reactions which maintain most of the steady states in the body are so complex, and are so peculiar to the living organism, that it has been suggested [...] that a specific designation for these states be employed – homeostasis" (Cannon 1929: 400).

Cannon already addresses the potential misinterpretation of stasis as being inflexible or even stagnating. Stasis also implies a certain condition, however, and in combination with the term 'homeo', meaning similarity, homeostasis is the concept of a system that is "to maintain uniformity" (Cannon 1929: 401). In this context, homeostasis is linked to the steady state concept (Lloyd et al. 2001), according to which such systems will remain constant despite influences from the external environment. Ashby (1952) calls this state a form of ultrastability, in which a system is able to change its internal structure in order to respond to the environment, causing the system to deal with external disturbances without compromising steadiness. Wiener (1948) formulized a form of feedback control that renders negative feedback as a critical source of reaction. Negative feedback is the response of a system to changes from a normal condition, from the steady state or the equilibrium, in order to move the system back to this normal condition. This is contrary to positive feedback, which would increase the departure from the normal condition. Therefore, in order for a system to be homeostatic, it needs negative feedback in order to react accordingly. A system is normally not able to achieve a stable homeostasis, but fluctuates around the equilibrium. Wiener (1948) expects oscillation and an eventual oversteering of the system. Such behavior can be traced back to the idea that negative feedback does not work in real-time and that any feedback comes with a certain time lag. A homeostat, as Ashby (1952) denotes a system in homeostasis, thus oscillates around the equilibrium, but keeps the system in an ultrastable condition.

Introducing big data to a homeostat implies an external disturbance that will probably lead to a massive deviation from the normal condition. A functioning homeostat changes its interior appropriately due to its ultrastability. Although the system returns to its equilibrium over time, the system is internally transformed and adapts to the new input. Interestingly, there is already some discussion of the idea that algorithms may be acting like homeostats has already entered discourse (Schwefel 1994). The idea implies that any algorithm-based system such as a modern organization tends to stabilize itself but will be transformed by big data. The logic of the homeostat emphasizes that big data may at first be a disturbance, but in the end will be used by the algorithmic system to return to the normal condition, especially as big data themselves, seen as an environmental force, do not have ultrastable features. Big data are constantly changing and transforming, and Ashby would, therefore, probably see big data as a source of variation, noise, and disturbance (Ashby 1952, 1956).

That may be a reason why Wiener declined the idea of homeostasis in society. He stated that "in connection with the effective amount of communal information, one of the most surprising facts about the body politic is its extreme lack of efficient homeostatic processes" (Wiener 1948: 185). In recent times of big data, the amount of information has drastically increased since Wiener's times, but there is still a lack of homeostatic processes. He anticipatively traced it back to the factor of numbers and size, which leads to "anti-homeostatic factors in society" (Wiener 1948: 187). He added that the "control of the means of communication is the most effective and most important" (Wiener 1948: 187). Thinking this further, due to the large size of big data, they can contribute to, but definitely *will influence*, smaller homeostats (any organization). Big data are not a homeostat on their own, however.

To follow those two concepts of first order cybernetics is the general idea behind second order cybernetics or cybernetics of cybernetics. The most important aspect of this new type of cybernetics is the renunciation of an objective reality, and consequently the impossibility of deriving an objective truth. Maturana (1970) and von Foerster (1979) connect the reasoning behind this argument to the observer of such a system. They stipulate that the claim of objectivity is in no way achievable due to the properties of an observer. Any observer will influence the observation to a certain degree. Von Foerster (2003) uses the following example to underline his argumentation.

> "... a brain is required to write a theory of a brain. From this follows that a theory of the brain, that has any aspirations for completeness, has to account for the writing of this theory. And even more fascinating, the writer of this theory has to account for her or himself. Translated into the domain of cybernetics; the cybernetician, by entering his own domain, has to account for his or her own activity. Cybernetics then becomes cybernetics of cybernetics, or *second-order cybernetics*" (von Foerster 2003: 289).

This argumentation reveals the importance of understanding the effect of an observer in any cybernetic system. This observer is not independent and is part of the observed system. This type of observer effect (Robins et al. 1996) is also known in quantum physics, where observing a quantum will change its properties, or where the observers influence their own observation by the mere act of observing (Heisenberg 1927). There is always an interaction between the observer and the observed. Observers influence the observed system with their eigenbehavior and the observers invent their environment (von Foerster 2003). Von Foerster paraphrases the effect as follows: "cognition → computing a reality" (2003: 215).

In the context of big data, the observers are not capable of separating themselves from big data at all, but now have enough information to compute a granular version of their reality. Within second order cybernetics, von Foerster (2003) covers the problem of memory. Memory, following his argumentation, is influenced by hindsight as well as foresight and, to make it even more difficult, the concept of big data is self-referential (Puschmann & Burgess 2014). Big data use data to generate new data in order to analyze data to generate even more data. In this recursive feedback loop, hindsight influences big data through experiences, and foresight is influenced by potentially desirable outcomes. Any observer will push any big data analysis into a new direction (willingly or unwillingly) and these new results will influence the existing observer or a new observer, in a different or the same way. Big data are part of a vicious cycle. Von Foerster summarizes it at the end of his chapter on constructing reality with the following claim: "reality = community" (2003: 227).

Although I suggest that this reality is a subjective one and not the objective reality, big data enable any form of community to construct their own. As noted earlier, big data lead to a data constructivism that exhibits a similar effect. The observer will influence the reality generated through big data and big data will carry this influence even further, creating a distinct eigenbehavior of the observer that spreads through big data. This eigenbehavior competes with other observers' eigenbehaviors, and will eventually lead to an eigenbehavior of the community. Such an understanding strongly supports the argument that big data are subjective (Boyd & Crawford 2012), and that raw data are an oxymoron (Gitelman 2013). Big data generate observations within the system observed by big data. On this basis, big data compute a subjective reality and *will not achieve* an objective reality or the objective truth, because big data are *both observer and observed object at the same time.*

Analyzing this selection of ideas in the theory of cybernetics from the perspective of big data reveals, above all, big data's inability of reaching the objective truth. Claims that big data will lead to an end of theory (Anderson 2008) can be disproven, and cybernetics implies that big data are bound to make understanding reality even more difficult. Cybernetics shows that an organization is obliged to deal with big data on its own in terms of variety, especially if big data influence a homeostatic organization and are influenced by the observer of big data, that is the organization itself. Big data and organizations interact in so many ways and so often that organizations influence big data and big data influence organizations. Any organization could use big data to *achieve ultrastability* within the modern turbulent environment. What ultrastability means, however, depends on the eigenbehavior of the organization and the resources it invests in dealing with the variety of responses.

2.3.2.2 Big Data in Systems Theory

The term 'systems theory' was coined by von Bertalanffy, although he more commonly addresses General Systems Theory (1968). According to the prefix 'general', the aim of the original theory is truly towards the full page as it is a type of meta-theory. Von Bertalanffy described it as a type of *weltanschauung* – world view (Pouvreau & Drack 2007). He did, however, identify a need for a systems approach for organizations in the 1920s (von Bertalanffy 1972) and explained this necessity, and the emergent overflowing of theories concerning organizations as follows:

> "[...] we are looking for another basic outlook on the world – *the world as organization*. Such a conception – if it can be substantiated – would indeed change the basic categories upon which scientific thought rests, and profoundly influence practical attitudes.
>
> This trend is marked by the emergence of a bundle of new disciplines such as cybernetics, information theory, general system theory, theories of games, of decisions, of queuing and others; in practical application, systems analysis, systems engineering, operations research, etc. They are different in basic assumptions, mathematical techniques and aims, and they are often unsatisfactory and sometimes contradictory. They agree, however, in being concerned, in one way or the other, with 'systems,' 'wholes' or 'organization'; and in their totality, they herald a new approach" (von Bertalanffy 1968: 187–188).

In his later research he narrowed down his general and somewhat holistic approach to shape the modern view of systems theory. He states that it is impossible for any person to grasp the objective reality or even the objective truth. It is only possible to mirror some aspects of this reality due to certain models (von Bertalanffy 1965). A system or organization can only be translated into models through a certain perspective, and such logic follows the concepts of second order cybernetics (von Foerster 1979) and radical constructivism (von Glasersfeld 1995).

On the premise of understanding organizations, system theorists have developed a variety of concepts that explain the behavior of organizations and the interaction of organizational members. There is, however, one concept that focuses purely on the description of an organization according to its input from the external environment and its output to, or the reactions of, the external environment. The inside of the system remains unknown and is described as a *black box* (Luhmann 1991): "The constitution and structure of the box are altogether irrelevant to the approach under consideration, which is purely external or phenomenological. In other words, only the behavior of the system will be accounted for" (Bunge 1963: 346). The observer is not able to see inside of the black box. The term 'black box' is also used in the context of programming. During what is called the 'black-box-test', a tester is to determine whether or not a piece of software is performing according to specifications, without knowledge of the software's inner workings. The opposite of this procedure is the 'white-box-test' and is conducted by the programmers involved in the development of the software. They possess knowledge about the code and are able to see the inside of the (white) box. The concept of the black box has become an integral part of today's programming culture.

These concepts from programming have an impact on the perception of big data. There is currently a tendency to put big data into a black box (Pasquale 2015), mostly because they seem too complicated to understand. Luhmann (1991) noted such complexity as another reason for black boxes. Nevertheless, the need to open the big data black box is urgent, especially as big data are currently capable of influencing every organization. Data are put into the black box that is big data, and completely new data potentially emerge from it. Especially in the context of organizations that are also potential black boxes (Sirmon et al. 2007), the interactions between both black boxes will seem difficult to follow. If big data construct a new type of reality and are on their way to becoming truly ubiquitous, seeing big data as a black box we will likewise place the entire system, and everything, inside of an enormous black box. With everything inside said box, everybody will also be inside which makes focusing on the input and output stimuli impossible. Although big data are currently observable to a certain degree as a black box, the interaction and especially the diffusion in, or fusion with, society will make these observations more and more complicated – at least to a certain degree. Ultimately, being within the big black box will create the necessity of *dealing with big data* as a white box.

The ability of an organization to self-organize (Nicolis & Prigogine 1977) and its capability of autopoesis (Maturana & Varela 1972) is connected to systems theory. Self-organization is the ability of an organization to achieve order out of itself. In a team, for example, self-organization is conducted by its team members. Self-organization will eventually lead to some form of spontaneous order (Kauffman 1993) and is presumably faster (Weick 1979). Autopoiesis is the potential of an organization to renew itself. Such an organization is autonomous from other organizations and capable of surviving due to its structure (Froese 2010). Both concepts assume that organizations or team members are independent and act freely, without influences from the outside. There are, however, restrictions to those concepts. Self-organization can be externally induced (Pongratz & Voß 1997, Stein 2000) and means that an organization can create structures that support self-organization and decrease centralized steering (Gomez & Probst 1980). Bounded autopoiesis (Scholz 2000) involves restricting the absolute autopoietic potential of an organization by putting in place a certain regulation rule that keeps the organization coherent.

Both concepts describe the behavior of big data quite well. Big data are in some form self-organized and autopoietic in one way or another. Data within big data interact with each other freely and will potentially find some form of order. In addition to that, big data constantly generate new data to achieve self-renewal. Although big data are capable of self-organization, they depend on a technological structure. People and machines generate data and interact with them, data will not create themselves without any external influence which renders self-organization impossible. Big data are also constantly renewing themselves. New data are added to the stockpile of data, but the blessing or the curse is that big data never forget (Solove 2011). This constitutes the need for some regulatory law that supports interaction with big data. Time, for example, may be a reasonable regulator, especially in the context of corporations. Corporations benefit greatly from big data, but the use of outdated data may cause serious harm to corporations. Although big data can self-organize and conduct autopoiesis, when interacting with organizations, the organizations have an interest in influencing the self-organization and autopoiesis in order to benefit from big data. Organizations want to *impact* and *interfere* with big data in order to *control* and *manipulate* the relevant portion of big data in their own interests.

Although there are many intersections with cybernetics, it is systems theory that underlines the claim for an inability to achieve objective truth. Nonetheless, big data represent a challenge to systems theory. Due to the ubiquitous amount of data, big data mimic a type of 'whole', but being so large will not fit into any kind of black box. Systems theory strengthens the idea that big data, as something that is everywhere, need to be researched and understood. Putting big data into a black box will not be sufficient, as the input and output of this big black box is also difficult to observe. Systems theory says that the internal structures of big data need to be observed in order to understand the effect of big data on any system. Systems theory also contributes to the idea that big data in their vastness cannot be understood, influenced or even changed by any system at an organizational level. Such a system can influence its perspective on big data. Big data that are relevant for organizations can be affected and interfered with. Such a task is also in the inherent interest of organizations: interfering in such a regulating way will help to harness the relevant portions for this system of big data. Such a system *acts proactively* and does not react to the output with which big data present it.

2.3.2.3 Big Data in Population Ecology Theory

The following theory focuses on the evolutionary approach to organizations. Rooted in the theory of evolution (Darwin 1859), population ecology theory considers the population of organizations and their battle for survival and, consequently, the survival of the fittest as a guiding rule. The survivability of an organization is only one aspect; more relevant are the evolutionary processes and the question of why some types of organization are more fit than others. Population ecology borrows the ideas of variation, selection, and retention or diffusion from biological evolution (Aldrich et al. 1984). Aldrich et al. (1984) describe these three stages as follows: *Variation* always takes place if a new organization is created, as this newly formed organization is influenced by existing ones and will blindly or purposefully vary from those organizations. *Selection* takes place because some organizations are more fit for the environment than others. Those organizations are able to acquire sufficient resources from the environment and will survive; other less fit organizations will have access to fewer resources and are bound to fail over time. This is a selection process that thins out the population and favors certain types of organization. *Retention* or *diffusion* concern the preservation of certain types of knowledge. This process not only focuses on the organizations as part of the population, but mainly targets the members of such organizations. Knowledge will be passed down in some way from existing members to new members, if it appears to contribute to the survivability of the organization. Aldrich et al. (1984) added the principle of struggle for existence, as organizations face fierce competition, although the time span of this competition as well the effects on an organization are more than decades long (Aldrich 1979, Hannan & Friedman 1977). The authors reason that organizations act as though they are in an evolutionary competition; observing this over such a long period of time, however, may be difficult. The concept introduces time as a factor for organizations, comparable to lifecycles (Hurst 1995). Organizations change over time in one way or another; organizations are created and will potentially die.

Big data can also be seen as a temporal construct. Big data and organizations are always in a relationship: one generates the other, which influences the first, and so on, until the starting point is no longer known. Big data, thus, underlie an evolutionary process. Data are generated and selected. If data are not objective, however, there will be errors in the analysis (Lazer et al. 2014). Big data mutate over time. On the one hand, mutations could be driven by data and lead to an autonomous evolution, which I propose as data-driven mutation. On the other hand, mutations could be described as organization-driven mutations. Both mutations are possible and probable; data-driven mutations, however, are currently popular and seem to be more adaptable to the environment (Provost & Fawcett 2013). Over time, there is the chance of a convergence of organizations as well as their fossilization. Data-driven mutations depend on big data as the source of selection, variation, and retention. As a result of big data, organizations become more specialized for a certain environment and make such organizations susceptible to environmental changes. Big data reinforce certain structures as big data also favor a standardization or normalization (Scholz 2015a). All of those factors add up to a hardening of the structures in organizations, because big data will act as Occam's razor (according to Wittgenstein (1922)). Here, data-driven mutations focus on plausible explanations while eliminating improbable ones. This is comparable to evolutionary degeneration or a reduction in variation in the population. Changing structures is difficult, and if reality is constructed around those structures, they turn into shackles.

As with biological evolution, organizations are sometimes not fast enough at adapting to new changes in the environment and are generally speaking not the fittest contestants in the population. In 1984, Hannan and Freeman developed the concept of *structural inertia* which refers to organizations that are not capable of understanding and predicting changes in the surrounding environment (Hannan & Freeman 1977), or unable to change internally due to a certain path dependence (Sydow et al. 2009). Such organizations are inert and will have lower potential survivability than more agile organizations. Although Hannan and Freeman (1984) propose that organizations rarely change, this claim is criticized by March (1981) as well as by the emergence of the research field of organizational change (Todnem By

2005). On top of that, the idea of differences in the capacity to change clarifies the reasons for the potential death of organizations (Freeman et al. 1983).

Introducing big data into the concept of organizational inertia will, at first, seem like a contribution. Big data support organizations with a variety of information, and may allow them to simulate or predict changes in their nature. Methods such as predictive analytics (Shmueli & Koppius 2011) allow organizations to prepare for changes in the environment and will lower structural inertia substantially but, having access to a huge stockpile of data concerning their environmental nature, will lead to more elaborate models and various different simulations. There is a natural complexity barrier (Gros 2012) within any of these big data models. One example is big data contributing to weather predictions (Hampton et al. 2013). Beyond a certain threshold, however, we see diminishing returns from using resources to deal with big data. It is possible to invest more resources into big data in order to get better weather predictions, but the amount of resources tied to improving weather predictions outweighs the outcome. Tying resources to predicting changes in nature will make organizations less agile, as organizations need more resources to foster their agility with big data. Big data will, therefore, make any organization more agile until a certain big data cap, beyond which any organizational resources bound to dealing with big data will decrease in agility and lead to an increase in structural inertia. Big data decreases inertia at first, but will eventually lead to an increase in inertia as shown in Figure 4.

Figure 4: Organizational Inertia and Big Data Cap

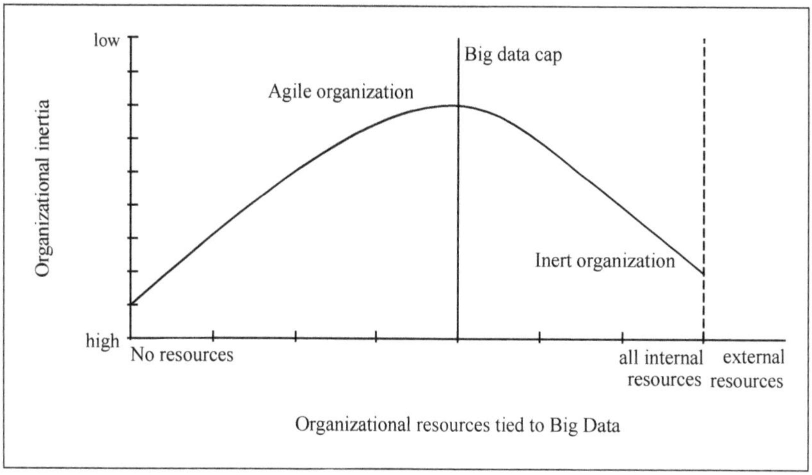

Population ecology theory supports several implications of big data and focuses predominately on the temporal perspective of big data. It seems that big data, at least in the short term, can have very beneficial effects on most organizations. Over time, however, these effects are subjugated to some form of diminishing return. As a matter of fact, big data may even make an organization less survivable when flowing unregulated into organizations. More and more resources are required to deal with big data which will generate a *data desirable structure*. Data-driven organizations are more specialized, but at the cost of their differentiation in structure and their potential agility. Population ecology theory identifies positive effects of big data to a certain point. At this tipping point, the potential benefits of big data are at their maximum. Above that threshold, however, big data become harmful to the survivability of an organization. Big data can be compared to oxygen: any living organism needs oxygen to survive, but an excess becomes toxic. Big data are essential for the survivability of any modern organization, but from a certain point onwards, big data are lethal for any modern organization.

Subsequently, the question arises: How can an organization discover this tipping point? It is essential to highlight that an organization cannot use big data for the discovery of such a tipping point, because any more resources allocated to big data will shift the organization further towards the big data cap or even worse beyond the big data cap. Therefore, the big data cap will be monitored by other means. This could be a monitoring by the respective experts within the organization and the observation of relevant indictors. Furthermore, the measurement does not require infinitesimal accurateness but ranges in certain intervals. The organization will already change in agility on the way to the big data cap, so the expert panel will perceive a diminishing return. The goal is to prevent the organization from reaching the big data cap and it will be sufficient to avoid a certain interval before this big data cap, in which agility is slowing down and/or decreasing.

2.3.2.4 Big Data in Complex Systems Theory

At the moment, many organizations are trying to solve problems using the classical playbook, and are focusing on simplification, predictability, equilibrium and linearity (Marion 1999). Barabási indicates the inadequacy of such an approach as follows: "As companies face an information explosion and an unprecedented need for flexibility in a rapidly changing marketplace, the corporate model is in the midst of a complete makeover" (2003: 201). Organizations need to move beyond reductionism (Barabási 2012) to a world where change is the new stability (Farjoun 2010). Complex systems theory focuses on unpredictability, non-equilibrium and non-linearity (Maguire et al. 2011).

The field of complex systems theory (or complexity theory) has a long history and is heavily influenced by cybernetics, systems theory, and evolutionary theory (Merali & Allen 2011). Although this theory seems like a loosely connected conglomeration of various concepts picked from different theories, the common notion or understanding of complex system theory is explained by Lissack, in that "within dynamic patterns there may be an underlying simplicity" (1999: 112). Scholz (2015b) points out that, as an organizational theory, complex systems theory has evolved from a "remarkable new vista" (Anderson 1999: 229) to "its time to change" (Andriani & McKelvey 2009: 1068) within recent years. Eoyang (2011: 320) even goes so far as to question everything we know: "Everything that supported stability and continuity of organization [is] compromised".

The field of complex systems theory is researched by numerous researchers, and there is a European school and a North American school, and many disciplines influence the field (Maguire 2011). For that reason, there is currently no concise definition available. There is, however, unanimity regarding the features of complex systems. Many researchers (e.g. McKelvey 2004, Sullivan & Daniels 2008, Maguire 2011) cite the description of Cilliers, concerning complex systems. He lists the following ten features:

1. "Complex systems consist of a large number of elements
2. A large number of elements are necessary, but not sufficient
3. The interaction is fairly rich, i.e. any element in the system influences, and is influenced by, quite a few other ones
4. The interactions are *non-linear*
5. The interactions usually have a fairly short range
6. There are loops in the interactions
7. Complex systems are usually open systems
8. Complex systems operate under conditions far from equilibrium
9. Complex systems have a history
10. Each element in the system is ignorant of the behaviour of the system as a whole, it responds only to information that is available to it locally" (Cilliers 1998: 3–4).

Those features are moving away from the general idea of reductionism and linearity. Their more complex direction is beneficial when it comes to big data. Big data consist of many elements and, even though the variety of elements may not be huge, their impact is ample. Big data, organizations, and especially the members of organizations, interact constantly, and this interaction is truly intensive. For Cilliers, the aspect of non-linearity is of utmost importance which is why he himself put it in italics. He explains that any large *and* linear organization will eventually split into similar but smaller organizations. Large organizations exist due to non-linearity. He reports that non-linearity "guarantees that small causes can have large results, and vice versa" (1998: 4) and implies phenomena like the butterfly effect (Lorenz 1963). Cilliers' fifth principle denotes the instance that interactions are of short range. To put it into context, big data may not directly influence a member in an organization, but the effects of big data are often the result of someone handling data within organizations (Rubinstein 2013). In addition to that, the system displays loops of interaction. Big data influence organizations as much as organizations influence big data, so there is a constant feedback loop in a complex system that is infused by big data.

Generally speaking, big data depend on the idea that an organization is an open system. Big data are big due to the idea that all data from every source are available.

Cilliers (1998) then compares equilibrium to the death of an organization. Although this may be a bit far-fetched, big data represent a constant source of disruption. Complex systems remember their history which is even more true for big data. Big data will remember everything that is collected about an organization. The information about such organizations is available for eternity. History may be ignored, but that decision would be made by organizations. Finally, Cilliers specifies that the elements in a system are nescient to the behavior of the whole system. This is reasonable because any element that understands the whole would inherit all the complexity of such a system. In terms of big data, no member of an organization will completely understand the wholeness of big data, or the impact of big data on an organization. Cilliers emphasizes a previous claim in a different context: only local (or organizationally relevant) big data are of interest to an organization, and organizations are only capable of dealing with those portions of big data. That means that within complex systems theory, big data cannot be completely grasped by any system.

On the premise that complex systems theory is rooted within the theories presented earlier, those concepts can be recognized within concepts of complex systems theory. They are expanded in certain ways and are part of advanced concepts tackling the same phenomena. As shown in table 8 and explained below, many of the concepts have a certain counterpart in complex systems theory. They may sometimes not precisely tackle the same phenomenon in an organization, but they are capable of describing the interaction between big data and an organization in more detail.

Table 8: Inclusion of Organizational Theory Streams in Complex Systems Theory

Theory	**Understanding Big Data**	**Expansion within Complex Systems Theory**
Cybernetics	Law of Requisite Variety	Complex Entropy
	Homeostatic	Homeodynamic
	Second Order Cybernetics	Third Order Cybernetics
Systems Theory	Black Box	Emergence
	Self-Organization	Self-Organized Criticality
	Autopoiesis	Fractals
Population Ecology Theory	Selection, Variation, Retention	Adaptation and Co-Evolution
	Organizational Inertia	System Fitness

Complex entropy. The first approach only vaguely fits the law of requisite variety, but it tackles the situation of how the system deals with variety in the environment. There is a metaphorical link between the law of requisite variety and complex entropy. The theory, however, moves away from the interaction between disturbance and response towards order and chaos (Whitfield 2005), or order and disorder (Morin 2008). Morin compares entropy to disorganization and uses negentropy (Brillouin 1953) with reorganization. He explains a certain paradox concerning entropy, in that the universe has a tendency to entropy (maximal disorder) on the basis of the second principle of thermodynamics, but that the universe also seeks to organize itself (maximal order). In the reductionist view, there is the assumption that we either observe order or observe disorder, but complex entropy allows for a system to go beyond such limitations. Morin claims that "for 'either/or' we substitute both 'neither/nor' and 'both/and'" (2008: 33). The paradigm of complexity for the author is not, therefore, the assumption that order and disorder are logically contradictory, but that order is linked to disorder. Order emerges from disorder and disorder is born in order. This view may not, however, be confused for a deterministic one, the question of who determines whom is irrelevant; what happens in the conjunction of order and disorder (Morin & Coppay 1983) is what matters. Complex systems theory is, therefore, concerned with the intersection at the edge of order and the edge of chaos (Waldrop 1993), thus tackling the first critical value (Bradford & Burke 2005) and the second critical value (Beinhocker 1997). Morin highlights the importance of linking apparently contradictory concepts when observing organizations:

> "If we think already that there are problems of irreducibility, of indeductibility, of complex relations between parts and whole, and if we think moreover that a system is a unit composed of different parts, one is obliged to unite the notion of unity and that of plurality or at least diversity. Then we realize that it is necessary to arrive at a logical complexity, because we should link concepts which normally repel each other logically, like unity and diversity. And even chance and necessity, disorder and order, need to be combined to conceive the genesis of physical organizations [...]" (Morin 2006: 9).

The conceptual grasp of this notion of big data is that they are simultaneously in a state of order and disorder. Any organization that tries to gravitate around the tipping point between the edge of order and chaos can use big data to achieve some sort of orbital stability. However big data are not inherently orderly or disorderly, but resemble a dynamic system that is influenced by people using big data, with big data changing themselves through algorithmic evolution. This means that either somebody external can enforce a predefined order, or big data can discover a spontaneous order. This order follows a certain rule and, as a result, will not be objective but subjective. Assuming that this order is subjective implies that disorder is also subjective. Consequently, both order and disorder will constantly compete against each other to achieve a certain type of order and disorder within big data. Big data within organizations will also follow such a process, trying to order and disorder the organization. If balancing between the edge of order and the edge of chaos already imposes a challenge in itself (Waldrop 1993, Marion 1999), however, having such strong and sometimes rampant force will bring about a destabilizing power that moves organizations far away from the 'sweet spot' shown in Figure 5.

Figure 5: Big Data as a Destabilizing Power for Order and *Disorder*

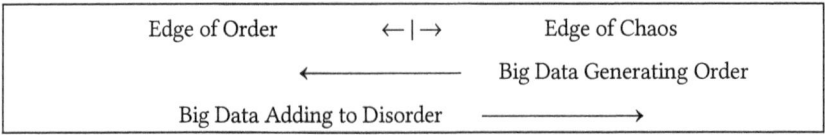

Edge of Order ←|→ Edge of Chaos

← Big Data Generating Order

Big Data Adding to Disorder →

In a recent study on information dynamics in social media, researchers discovered that within the diffusion of information, there is a phenomenon they called the "order-disorder-transition" (Borge-Holthoefer et al. 2016: 6). They propose that, if information is spread far enough, any information network will transform from centralized to decentralized, and consequently, shift from order to disorder. The researchers assume that such social networks are not controlled or steered by an external force. Big data, therefore, possess the ability to add disorder in big data and within an organization. Big data are able to move an organization away from the edge of order. Rätsch (2015) discusses the potential of big data to prevent innovation and lead to an organizational stalemate. Picking up his argument, organizations using big data for the sake of order become more average. In a sense, being average is not a bad thing (Scholz 2013a), but it will lower disorder and eliminate variety in organizations. Furthermore, such developments reinforce themselves, if big data suggest that a certain structure is beneficial and big data obtain beneficial results, the organization will be forced to follow its path, structures will become shackles (Scholz 2015a), thus generating a data-driven structure. Big data will generate order and move organizations away from the edge of chaos. Using big data deliberately is a premise to keep an organization in orbit around the joint between the edge of order and the edge of chaos.

Homeodynamic. If an organization can only gravitate around this point of order and chaos, it will cope with ordering and disordering forces, but achieving a form of homeostatic steady state appears impossible (Lloyd et al. 2001). To conquer these obstacles, Yates (1994) developed the concept of homeodynamics, a concept that has some similarities to dissipative structure (Prigogine & Stengers 1984) and homeokinetics (Soodak & Iberall 1978). Trzebski (1994) describes the main difference between the concepts as follows:

> "Homeostasis (is) state oriented homeostatic steady state, stability close to equilibrium, Program (set point)-driven system. Homeodynamics (are) rate-oriented homeodynamic stability, not very far from equilibrium, fluctuating and oscillating or close to 1/f noise informationally, not fixed program-driven systems with easy generation of new activity patterns" (Trzebski 1994: 111).

Organizations try to achieve a certain type of homeodynamic stability (Scholz & Reichstein 2015): stability here does not refer to a steady state, but rather to the actual idea of stability. If organizations deal with external changes and are able to self-organize internally, stability becomes a dynamic concept. Farjoun (2010: 203) describes stability, and reliability, as follows: "In their more dynamic sense they can also be viewed as long term efficiency and robustness against failure and persistent perturbations [...] and they therefore require variation to sustain". He, therefore, contradicts the idea of a steady state and implicitly proposes a more homeodynamic view of organizations.

Talking about big data and a steady state or equilibrium also sounds paradoxical. Big data are, above all, dynamic and largely imbalanced. The only steady thing about big data is their exponential growth. But even the exponential will change one day, due to the complexity barrier and diminishing returns (Gros 2012). For big data, the steady state is growth. For an organization this means that it copes with the influence of a growing amount of data. In order to achieve some form of homeodynamic stability the organization changes so as to oscillate near the equilibrium, transforming or filtering the increasing big data noise into something useful. Homeodynamics are, therefore, achieved by dampening or even amplifying the effects of big data in a way that causes organizations to achieve a (temporary) homeodynamic balance.

Third order cybernetics. In cybernetics there is already a differentiation between first order and second order cybernetics. Recent years have seen discussions about a third order cybernetics (e.g. Boxer & Kenny 1990, Johannessen & Hauan 1994, von Foerster 2003). The discourse about third order cybernetics is closely entangled with the emergence of virtual reality concepts such as the internet and other media (Kenny 2009). Although the singular term virtual reality is defined more loosely than its plural virtual realities such as World of Warcraft or other video games, the general idea is that people face life or reality in a virtual environment. Big data contribute massively to the data-constructed reality that is happening in this virtual reality. Kenny (2009) explains that nothing is real and that it no longer seems possible to identify the observer. He also says that it is questionable whether the observer has seen anybody or just observes phenomena derived from data. Big data allow for more granular data to be gathered about individuals, but always correlate to many other people (Tene & Polonetsky 2013). In order to generate Amazon recommendations, for example, the system takes a user's recent purchase and compares it to those of millions of other customers and the products they ended up buying. Users simply mimic the behavior of others and others mimic theirs. So, is the system really observing individuals or just a mass of people? Big data make this differentiation more difficult. Kenny (2009) asks if anybody even knows their observer. We are now living in a world of mass surveillance and 24/7 connectivity. We are observed at all times, but do not know by whom. Similar to the post-panopticon (Baumann 2000), we are well aware that observation takes place; but that is all.

Kenny (2009) proposes third order cybernetics in the sense of big data, an expansion already described by the example of big data. It is essential, however, to emphasize that big data engulf the idea of an observer. Everybody observes everybody and everybody is observed by everybody. Big data are also never real, neither in actual nor in virtual reality. Big data construct a subjective reality in both real and virtual worlds. These two constructed realities are not necessarily identical or comparable and may resemble the idea of the presentation of self, as proposed by Goffmann (1959).

Emergence. Black boxes are evidently incapable of grasping big data, as big data are something completely new and stimulate a massive increase of thinking. Big data cannot be put into any existing black boxes. This newness is the source of new knowledge, new ideas, and change in a system exposed to big data. Consequently, any system connected to and influenced by big data will change over time and generate new patterns. Some form of emergence is inherent. Emergence is not a new concept and in terms of social systems can be attributed to Durkheim (Bellah 1959). Sawyer (2002) analyzes Durkheim's contribution to emergence research in social systems. From the interaction of individuals, Sawyer explains, some form of patterns emerges at the social level. It is what McKelvey (2016) describes as the bottom-up emergence of strategies (Mintzberg & McHugh 1985), ideas (Hamel 1998), networks (Feldman & Francis 2004), groups (Barry & Stewart 1997), hierarchies (Illinitch et al. 1996), or emergent innovations (Oster 2009). Emergence is a phenomenon that, especially in complex systems (Holland 1995), has a strong impact at the higher level, as the following quote explains:

> "There can be no sociology unless societies exist, and ... societies cannot exist if there are only individuals" (Durkheim 1897/1951: 38).

The quote applies to big data as well: there can be no big data unless data exist, and data cannot exist if there are only *datum points*. Out of big data emerge big data and big data generate big data. It may be unclear, however, due to its complexity, what exactly emerges from big data. Big data have a generative capacity. Similar to the concept of generative grammar (Chomsky 1965), according to which grammar as a set of rules generates language, it is known that the internet is also capable of being generative (Zittrain 2006). New patterns will be able to emerge from big data as form of informational grammar.

Self-organized criticality. Such emergence also needs a form of self-organization which is a concept popular in complex systems theory (e.g. Kauffman 1993, Krugman 1996). Bak et al. (1988) extended the concept of self-organization with the idea of self-organized criticality. In a later book, Bak (1996) describes the concept as follows: "I will argue that complex behavior in nature reflects the tendency of large systems with many components to evolve into a poised, 'critical' state, way out of balance, where minor disturbances may lead to events called avalanches, of all sizes. [...] The state is established solely because of the dynamical interactions among individual elements of the system: the critical state is self-organized" (Bak 1996: 1–2). He clarifies the idea employing the example of a sand pile. Adding sand to a sand pile is understandable and observable at first, but adding sand will make the system grow, causing it to eventually establish a dynamic of its own. Avalanches may occur, when more sand is added to the sand pile. Bak concludes from this idea that, although individual actions are understandable, they become complex when embedded in a bigger environment (e.g. nature). Individual and local interactions are still possible in such a dynamic system. Those small interactions, however, can grow into big avalanches. As the system grows, these effects can become stronger, a phenomenon known as the butterfly effect (Lorenz 1963).

Big data as a system are huge and full of interactions. However, all interactions take place independently and autonomously. Local interactions between data take place everywhere and at any time. Big data are submitted to self-organized criticality. One example is high-frequency trading (Buchanan 2015) and the flash crash (Kirilenko et al. 2014) which was caused by a small error in one algorithm, and led to a massive stock market crash in 2010. Although the error was quickly discovered and corrected, the market did not rebound completely. Ultimately, one algorithm interacted with another algorithm and these local interactions disrupted the entire system. One of the advantages of big data is that they are fast and are found almost in real-time which means that there are many interactions between data within big data. Big data in an organization are also highly self-organized, they interact with members of organizations on a regular basis, and the sand pile of big data in the organization will grow over time within an organization. Self-organized criticality will also grow, rendering big data within organizations highly complex.

Fractals. Autopoiesis is often linked to the idea of self-similarity (Andersen 1994) and seeks from this similarity the potential for something to renew itself. Maturana (1987) criticizes the application of autopoietic behavior within social systems, but is intrigued by the idea that there is some form of governing rule. Some researchers (e.g. Eoyang 2011, Scholz 2015b) explain self-similarity within organizations using the concept of fractals (Mandelbrot 1977) and describe a geometric shape that, if split into parts, seems like a smaller copy of the whole. The most popular example of a fractal is the snowflake. Falconer (1997) even reports the possibility of generating a fractal iteratively through a non-linear equation. In organizations, fractals are used as a metaphor (Eoyang 2011) and refer to knowledge (Nonaka & Takeuchi 1995), ideas and innovation (Zimmermann, & Hurst 1993), and corporate identity (Bouchikhi & Kimberly 2003). Knowledge, ideas, and identity are quasi-fractals at the individual level and the organizational level. They have a certain self-similarity and are essential to the individual-organizational fit.

Fractals within big data are also more of a metaphorical concept. Big data at the societal level can be similar to those at the individual level. From a statistical perspective, there is a certain inherent self-similarity. Individual data are aggregated at a societal level and big data are consequently fractals, big data can become fractals due to this idea. Scholz (2015b) proposes the following argument, that if an organization uses a normal (Gaussian) distribution and, therefore, focuses on the average, fractals enforce a more centralistic view. The majority decides what those fractals look like. The minority accepts and adjusts their fractals. Big data have the capability of reinforcing such behavior. Big data are, therefore, not fractals, big data *create* fractals on the basis of a certain governing rule and which makes them *bounded* fractals.

Adaptation and co-evolution. Complex systems theory, in general, is heavily influenced by evolution theory. The main premise of selection, variation, and retention is dominant in complex systems theory (e.g. Holland 1995). The main change is the understanding of evolution as a dynamic system. Organizations constantly adapt to internal and external changes (Siggelkow 2002) and establish an environment of co-evolution (Rindova & Kotha 2001). Individuals in organizations also influence adaptation and co-evolution (Stacey 2001).

Big data and organizations dynamically adapt to each other and co-evolve. Such behavior can potentially cause a rat-race and mimic the behavior noted in the red-queen hypothesis (van Valen 1973). Both systems constantly try to improve their survivability in competition with other opposing systems. Van Valen derived the term from Alice in Wonderland: "Now, *here*, you see, it takes all the running you can do, to keep in the same place" (Carroll 1991). Big data are ubiquitous and any organization has access to a vastness of data. There is an impending need to use big data and many organizations use them simply because other organizations do. Such an evolutionary race is already happening, and organizations are running in a certain direction unaware of whether it is the direction with the *highest* survivability. This concept is linked to self-organized criticality (Adami 1995). The current evolutionary path could lead to an evolutionary dead end (Takebayashi & Morrell 2001).

System fitness. Finally, there is the concept of organizational inertia and, in the context of complex systems theory, the ability or inability to react quickly to internal and external challenges is often referred to as the fitness of a system (Anderson et al. 1999). Evolutionary adaptation could eventually lead to an evolutionary dead end, but some populations are capable of changing direction completely. Such populations are fitter and can change dynamically according to changes in the landscape. Kauffman (1995) borrowed the term "fitness landscape" from Wright (1932) and theorized that some populations are more adaptable than others. This form of fitness is visualized as height in this landscape (Provine 1986). The higher a population, the fitter it is. Over time, such landscapes can change dynamically and something that was defined as being fit may become less fit.

After exemplifying the expansions of big data within complex systems theory as depicted in Table 8, this dynamic approach is especially relevant for big data in organizations. It may seem less cost-efficient to focus on one type of fitness with respect to big data. That could mean using only one type of big data analysis for all big data problems. Within a static environment, however, such an approach is evolutionally correct and will result in the fittest solution. Specialization trumps generalization in this fitness landscape, but within a dynamic environment the fitness landscape is dynamic and changes constantly. One solution may sometimes help, but may otherwise be pointless. An organization is, therefore, able to conduct a variety of big data analysis. However, organizations are also able to identify a fitting analysis for current respective evolutionary obstacles. Generalization trumps specialization in this fitness landscape. An organization that stays *homeodynamically agile* will be fitter than a highly specialized one.

Big data can be grasped by complex systems and, above all, reveal the need to deal with big data within an organization. Being complex, however, does not equate to the idea of making something complicated and does not denote a decision between reductionism or holism. As stated by Morin (2008: 56): "Complex thought does not all reject clarity, order, or determinism." Complex systems can be governed by simple rules (Eoyang 2007, Sull & Eisenhardt 2012) even if the system is dynamic and flexible (Falconer 2002). I, therefore, follow the proposition of Farjoun, of moving beyond dualism towards a type of duality: "Duality resembles dualism in that it retains the idea of two essential elements, but it views them as interdependent, rather than separate and opposed" (2010: 203). This aligns with the demand by Morin (2008: 33) to "substitute either/or for both/and". There is a need to balance "both stability and flexibility, both continuity and disruption, both ties to the old and stretches to the new" (Eoyang 2011: 326). In summary, big data within an organization will tackle the order and disorder with drastic measures and will continuously influence organizations. Organizations, therefore, will find ways of dealing with big data. In order to gain a competitive advantage and an evolutionary lead from the use of big data, it is necessary for an organization to achieve a *homeodynamic stability* and stay *homeodynamically agile* in the context of big data within organizations.

2.4 Big Data at the Human (Resource) Level

2.4.1 Current Status of Big Data in Human Resource Management

Big data will have an extensive impact at the social level, the organizational level, and the individual level. Especially within an economic organization the effect of big data will transform the way people are working. Initially big data will change the way the HR department is working, and only after this change will the effects of big data influence every employee within the organization. Consequently, the impact at the human resource level precedes the impact at the human level, although the human resource level already comprises an influence at the human level – within the HR department.

Nevertheless, big data in HRM is currently underresearched (e.g. Angrave et al. 2016, George et al. 2014, Huselid 2015) and, although, big data will influence human relations (Harvard Business Review 2013) the current discussion is driven by practitioners rather than researchers. The relation between HRM and big data is quite interesting as HRM holds the competence to support human actors as well as the strategic potential to implement big data into organizations, although its technological competencies are currently underdeveloped (Stone et al. 2015).

In the context of big data and HRM probably the most cited case is that of Moneyball (Lewis 2004). The author discusses Billy Beane and his experience as the general manager of the baseball team "Oakland Athletics". The book represents a fitting example of big data in HR, because the players are the most valuable asset a sports team holds. The team's narrow budget forced the manager to search for different ways of acquiring talent. Using and analyzing big data, he managed to form a team that was unusual, but competitive and highly successful. He discovered new indicators to evaluate the performance of players providing a competitive advantage towards other teams. Today's baseball teams employ so-called sabermetricians in order to level the playing field by analyzing empirical data (Baumer & Zimbalist 2014). Beane's competitive advantage is now available to every team in the league. Those approaches to the use of statistics have become popular, especially in team sports – for instance, in ice hockey (Mason & Foster 2007), basketball (Oliver 2004), and soccer (Anderson & Sally 2013) to name a few. This example reveals a strong focus on strategic HRM, consequently, there is a link between big data and strategic HRM (Angrave et al. 2016).

The field of strategic HRM emerged as a research stream in HRM in 1984 and can be traced back to the research by Beer et al. (1984) and Fombrun et al. (1984). Over the time the definition of strategic HRM changed and this progress is described in an article by Kaufman (2015). In his review article about the evolution of the term strategic HRM, Kaufman summarizes the central elements of strategic HRM as follows:

> "HRM as the people management component of organizations, a holistic system's view of individual HRM structures and practices, a strategic perspective on how the HRM system can best promote organizational objectives, HRM system alignment with organizational strategy and integration of practices within the system, and emphasis on the long-run benefits of a human capital/high-commitment HRM system" (2015: 396).

This synopsis highlights the integral role HRM plays in organizations and the general strategy. There is a fit between the work of the HR department and the strategy implementation within the organization; consequently, strategic HRM contributes towards the competitive advantage of an organization (Becker & Huselid 2006). Therefore, if the HR department is a source for strategic decisions and, by that, contributes to the competitive advantage, this HR department needs to have a high differentiation in its architecture (Lepak & Snell 1999). Furthermore, Becker & Huselid (2006) mention that in order to contribute towards the strategic direction and the potential competitive advantage, the HRM focuses on its system rather than operational tasks. The strategic goal of HRM is to contribute to a sustainable competitive advantage. However, the focus will not purely lie at the organizational level but also at the individual level (Gerhart 2005). Strategic HRM is, therefore, a link between the strategic direction of the organization and the impact of such strategy at the individual level.

Due to the reason that big data influence the organization extensively, strategic HRM will deal with those changes in a strategic way to generate a competitive advantage out of big data. It is important to highlight that in this case, the competitive advantage is generated by combining people with big data. Big data aligned with the current digitization resemble a paper by Lepak and Snell (1998) talking about the virtual HR department. Big data enable the HR department to have access to all the relevant information as well as communicate with every employee everywhere. Interestingly, they highlight that "perhaps the most dramatic impact of IT on structural integration within HR is its transformational role" (Lepak & Snell 1998: 220). Derived from that, big data will transform HRM – HRM, however, will exploit the technological potential of big data, in order to do its work in a more flexible, more dynamic, and more responsive way. Big data enable the HRM to strategically realign itself, in order to transform the working environment for its employees. The focus shifts from operational tasks towards a more strategically oriented management of the organization and the relationship between people and big data. Technology is seen as a catalyst for the change of the HRM function (e.g. Parry 2014) and, therefore, big data enable HRM to focus on the strategic perspective and to create an environment for the employees that may lead to a competitive advantage.

Big data could lead to freeing up resources in the HR department that are currently used to do operational tasks. Tasks which can, potentially, be automated and would enable the HR department to focus more on strategic work. However, the current situation of big data in HRM is quite different. Although it seems obvious that big data will, predominately, require a strategic HRM and the Moneyball example highlights this necessity, current applications derived from Moneyball are on an operational level in areas like recruitment (gild 2013), talent management (Bovis et al. 2012), job performance (Armstrong 2012), and data-driven decision-making (Guszcza et al. 2013). Table 9 depicts further opportunities for the use of big data especially in the field of recruitment. Big data may aid in the search for candidates and provide insights into the recruiting process. The use of big data supplies additional benefits for workforce planning and the talent management of employees. However, big data are not seen from a strategic perspective in HRM. It can be stated that the strategic HRM perspective suffers neglect when it comes to the application of big data at the human (resource) level.

Table 9: Examples of Big Data in Human Resource Management Practice

Categorization by Armstrong (2014)	Operational HRM	Strategic HRM
People resourcing	**Workforce planning:** • Employee development planning (dm) • Labor management (Blue Cross Blue Shield, LBHF)	**Workforce planning:** • Demographic risk management (Deutsche Bahn) • Strategic personnel planning (EnBW) • Strategic workforce planning (ÖBB, Techniker Krankenkasse, Commerzbank) • Strategic scenario planning (REWE, Lufthansa, Bayer MaterialScience) • Scalable workforce planning (AccentCare)
	Recruiting: • Recruiting with focus on niche roles (Tripadvisor) • Recruiting with focus on experience-passion-fitness (fitbit) • Candidate engagement (Rapid7) • Candidate management (Recurly) • Intelligence & insight into candidates (red hat) • Candidate search efforts (StrongView) • Recruiting hidden talents (Taboola) • Recruiting and time-to-hire (Fitness First) • Candidate communication and employer-branding (Fitch Ratings) • Web-based recruitment management (Carillion) • Recruiting non-exempt employees (CARQUEST) • Hiring process system (Apollo Group)	**Recruiting:** • Establishing and maintaining talent pool (Gainsight)

Categorization by Armstrong (2014)	Operational HRM	Strategic HRM
	Talent management: • HR life cycle with focus on talents (Avaya) • Talent management system (Motorola, Nationwide, WakeMed)	**Talent management:** • Talent acquisition strategy (Advance Auto Parts)
Learning and development	**Human resource development:** • Learning management system (UAP) • Learning content management system (Potash) • Employee development program (NYC: Department of Education)	**Human resource development:** • Improving training attrition (JetBlue)
Performance and reward	**Performance measurement:** • Standardization of specific measures (Evonik)	
	Compensation and incentives: • Incentive management (Financial Service Company) • Compensation management (Exelon, Scotiabank)	
Employee relations	**Employee engagement:** • Employee engagement program (FRHI Hotels)	**Employee engagement:** • Change in morality and attrition (Nationwide Brokerage Solutions)
	Onboarding: • Onboarding process (H&R Block)	
Employee well-being	**Healthcare:** • Evaluation of healthcare costs (Wegmans)	

(Cases from Blue Yonder, Dynaplan, glid, Google, IBM Kenexa, PeopleFluent)

Nevertheless, neglecting the strategic HRM of big data will not be productive and could even be harmful (e.g. Peck 2013). Consequently, there is a certain research gap in the field of big data in HRM as shown in Table 10. The research in HRM currently struggles to transfer the basic research into some form of applied research and by that widening the research to practice gap (Huselid 2011). However, at the same time Anderson (2008) states that big data will work without any theory and, by that, proposing the sufficiency of data-driven applied research. At this point in time, data-driven applied research is dominating the field and there is no known theory-driven applied research. This becomes explicable in the fact that data-driven applied research is quicker than theory-driven applied research, especially as big data are still not sufficiently understood theoretically. Suggesting that theories are no longer necessary, and that with enough data valid results are possible regardless of theory, sounds compelling to many, especially to corporations. Therefore, data-driven applied research has a significant head start compared to theory-driven research. And it also explains the focus on operational applications.

Table 10: Hermeneutical Observation of Big Data in HRM

		Way of Gaining Insights	
		Theory-Driven	Data-Driven
Research Focus	Applied Research	?	See Table 9
	Basic Research	Scholz 2015a	Anderson 2008

Both approaches appear contradictory. Furthermore, the current dominance of data-driven applied research neglects strategic HRM and purely focuses on operational HRM. Therefore, every application of big data in HRM lacks a strategic fit (Scholz, C. 1987) towards the organization in any aspect. Consequently, there is currently in most organizations no link between HRM and big data strategies; however, such a link is essential to utilize all resources within an organization (e.g. Scholz, C. 2014a). Big data are decoupled from the strategic HRM, however, influence the operational HRM due to data-driven applications. Such applications lead to data-driven decisions and, therefore, are indirectly influencing the strategic HRM. Big data and strategic HRM cannot be separated and are highly linked, and whilst at the moment big data determine the work of strategic HRM, the task of strategic HRM is to strategically manage big data in HRM. The usefulness of such a function is highly debated (e.g. Cappelli 2015, Charan et al. 2015).

The current imbalance creates ground for the ongoing turf war within HRM. HRM is already facing an existential crisis (Ulrich et al. 2013). Data-driven applications take over several core fields of HRM and HRM is at the moment neglecting the chance to focus on the task of strategic HRM. Consequently, the role of HRM is shifting and its path is unclear.

2.4.2 Classification of Views

For the sake of terminological division of the two approaches, those supporting the data-driven approach shall be called *anti-guessworkers* and those following the theory-driven approach to be called *neo-luddites*. The two terms are not intended to be judgmental, but they describe a certain behavior or attitude of the groups.

Supporters of the data-driven approach do so in order to eliminate the "guesswork" (Evolv 2013) involved in HRM. Those people characterize the HR department as "being touchy-feely, but in the age of big data, it's becoming a bit more cold and analytical" (Walker 2012a). Block (as cited in Walker 2012b) even goes so far as to state that "software will supplement, if not supplant, many of the personnel decisions long made by instinct and intuition." Big data have finally led to HRM analyzing at least some data. By using distinct measures and metrics, it is possible to lower the employees' sick time, increase retention, lower attrition, and optimize payment (Walker 2012b). Another example is Google's Project Oxygen (Bryant, A. 2011): by analyzing data such as performance reviews and surveys, a team derived rules of leadership, such as being a good coach or having a clear vision and strategy for the team. Others have discovered that there is a correlation between the browser on an employee's computer and their performance (Economist 2013). These examples show that big data can tackle some important questions; it is essential, however, to select the right sense-making metrics (Bladt & Filbin 2013). A popular example is the aforementioned Moneyball example (Lewis 2004), a seemingly purely data-driven approach that led to the major success of the Oakland Athletics.

Contrary to the view on big data in HRM, the neo-luddites claim that HRM can work professionally without having to take such an intensively data-driven approach. The term luddite is derived from the anti-technological-progress movement in the beginnings of the industrial revolution (Baggaley 2010) and is picked up in recent years in a populist fashion in terms of automation, claiming for the "race against the machine" (Brynjolfsson & McAfee 2011) to be common. These neo-luddites, in the context of HRM, are especially offended by the fact that current HRM only uses guesswork instead of any distinct analytics (Lay 2012). They even accuse big data of "dehumanizing human resources" (Cukier 2013) in claiming that along the road of big data, humans will turn into nothing more than resources (Graham 2013). Decisions and processes will be outsourced to big data and the analysts. They also question whether or not the behavior and actions of employees can be sufficiently collected as data (Williams 2013). Even if the existing data are good, they will not necessarily lead to good decisions (Shah et al. 2012). The neo-luddites especially address the privacy aspects of big data in the light of the global surveillance disclosures in 2013, building up resistance towards the use of big data and reinforcing the HR-IT barrier. Claims that all data will be collected (Richtel 2013) and focusing on the data exhaust or the digital footprint (data that we leave behind) increase skepticism against big data. While vast amounts of data may be of interest for HRM, they make the employee transparent, and damage the trust between HRM and employees (Scholz, C. 2014b, Scholz, C. 2016), thus, therefore, destroying the key to high employee morale (Graham 2013). The neo-luddites see parallels to Taylorism and call this new form of workplace surveillance "new digital Taylorism" (Parenti 2001: 26) or Taylorism 2.0 (deWinter et al. 2014). Some authors even go so far as to argue that data will be the resource of any knowledge and cognition (Anderson 2008). Big data create a *holistic* picture of employees and are already being used in determining an individual's use to an organization without that person having a chance to justify themselves – a procedure with striking resemblance to "Der Process" (Kafka 1925) in which the protagonist is prosecuted for a crime: although the crime he is charged with is unknown, the jury receives details of his life and consequently finds something incriminatory.

Even though both sides obviously exaggerate their claims and, apparently, strongly oppose each other, both contribute towards the *erosion* of the HRM function in an organization (Ulrich et al. 2013, Cappelli 2015, Charan et al. 2015, Stone et al. 2015). On the one hand, the anti-guessworkers are implementing data-driven structures that will eventually lead to the *obsolescence* of the HR department. Why does an organization need such a department, if everything it does can be done by a data-driven application, especially, if those applications are faster and apparently more precise (e.g. Brynjolfsson et al. 2011, Feffer 2015)? On the other hand, the neo-luddites contribute to a strengthening of the current HR-IT-barrier. The HR-IT-barrier describes the complications between the HR department and the IT department to communicate properly with each other. The HR department loses its connection to the organization and can no longer contribute to it. In these times of digitization, in particular, the HR department is also transforming into a more digitized function. Not using those new tools as well as limiting all dispositions of them leads to a decrease in usefulness of the organization. For this reason, both approaches seem to be insufficient in trying to grasp the relationship between big data and HRM.

2.4.3 Augmentation as an Alternative Path

The anti-guessworkers accuse the neo-luddites of being too human while on the other side of the spectrum, the neo-luddites blame the anti-guessworkers for being too mechanical. Both groups seem like specialists in their respective fields. Big data at the human level are simultaneously mechanical and human. The individuals in consideration are not a string of zeroes and ones but people and people following their instincts may potentially be beneficial or harmful for an organization. The emphasis here, lies on the word *may* as it reveals a certain uncertainty. Whether big data are good or evil is debatable, yet the question is far from constructive. Big data are here to stay and big data are shaping the reality of people. Big data are entangled with their lives, especially with their working lives.

Rather than denying the advantages of either side, I propose a different approach. Following the logic of duality, this approach shall be called *augmentation* approach. Augmentation derives from the Latin word *augmentare* and means to gain, add, foster, or increase. One thing is augmented by another thing and becomes more, bigger, or better. In today's world (and fitting for the argument of this thesis), the idea of "augmented reality" (Azuma 1997) is a very popular one. This technology mediates the view of reality by adding an additional layer of vision to it. The perception or view of a person is enhanced by this additional layer (Graham et al. 2012). Users of augmented reality receive more information which enables them to improve their work. Using this as an analogy, augmentation seems to fit the case of HRM.

Augmentation also describes a certain direction of use. The human actor is augmented by technology to become *better* which does not exclude a data-driven approach rather than narrowing it down. A human is responsible for the decisions made and is only augmented by big data in order to make the best decision possible in a distinct case. There is room to be humane in certain cases, but also the potential of supporting decisions with big data. Big data and HRM can work together and their collaboration can be superior to either one working alone. The superiority of such collaboration has been proven in chess, in which the most successful combination is human and machine together (Kelly 2014, Ford 2015).

This augmentation also allows for big data to connect at the human level. People are capable of using big data for their purposes and are able to utilize and harness their potential. This is relevant to the use of big data at an individual level, but also on an organizational one. In the previous chapters it has become evident that big data are omnipresent and surround people and society, but both worlds seem to be separate from each other. Big data and society, however, are closely entangled and interact extensively. Implementing such an augmentation makes big data visible and usable for the HR department and, ultimately, for individual employees.

Interestingly, although Moneyball is often seen as an example of the superiority of big data in HRM, the current development shows a different path. Fears have developed that the big staff of coaches, scouts and others may become obsolete (Kim 2014). However, the use of the Moneyball principle had different outcomes. The Danish football team FC Midtjylland uses big data for their work (Biermann 2015) and has recently won the Danish championship for the first time. The result of this approach is that people view football differently. Big data have opened up new possibilities. Biased decisions by coaches or staff are debunked through big data which enables people to bring the best team together. Big data help to buy the best players for any budget, but money "can't buy team spirit" (Thomas & Wasmund 2011: 286). Big data augment and support people within organizations in making better decisions, but big data do not make people redundant in the process of decision-making. Biermann concludes that the competitive advantage of FC Midtjylland is not a result of big data but of the "synthesis of cold analysis and heart" (2015: 96). It seems that big data, at least in this case, are not the source of competitive advantage (contrary to the Moneyball case), but that the people are the competitive advantage. This may be even more true in a world where everybody has easy access to big data. Such a task sounds similar to the prime goal of HRM: *making people better*!

3. Research Framework

3.1 Mental Model

After introducing the term 'big data' and defining it on the basis of several theories, the logical next step is the analysis of this reasoning with regards to organizations, as well as an estimation of the HR department's potential future role. In a first step, I will propose my mental model for this new form of organization and the new role of the HR department. This model serves the purpose of outlining the reasoning behind my framework. Wittgenstein describes the need for such a model as follows:

> "We make to ourselves pictures of facts. The picture presents the facts in logical space, the existence and non-existence of atomic facts. The picture is a model of reality" (Wittgenstein 1922: 28).

The term 'mental model', in this context, was first used by Craik (1943) who described it as the ability of an individual to use external input from alternative models and derive the best alternative, resulting in one concise model that can be presented to other people. Johnson-Laird (2004) describes the cognitive map (Tolman 1948) and the work of Peirce (1958) as precursors to the mental model. However, the term 'mental model' gained popularity through Forrester and his definition:

> "The mental image of the world around you which you carry in your head is a model. One does not have a city or a government or a country in his head. He has only selected concepts and relationships which he uses to represent the real system. A mental image is a model. All of our decisions are taken on the basis of models. All of our laws are passed on the basis of models. All executive actions are taken on the basis of models. The question is not to use or ignore models. The question is only a choice among alternative models" (1971: 112).

Forrester explains that a mental model is not precise, but fuzzy, and is not complete but fragmentary. This is especially the case in social systems which is why knowledge about social systems may be insufficient. Forrester claims that "we do know enough to make useful models of social systems" (1971: 111). Consequently, the function of a mental model is to parallelize the thinking of individuals in order to achieve learning and knowledge that reach across individuals and do not solely exist within the individual's mind (Senge 1990).

The interaction between people and big data will particularly be in dire need of mental models, in order to generate alternatives, be able to gain a better understanding and improve knowledge about this complex interaction. I will develop the mental model of big data within organizations, as shown in Figure 6.

Figure 6: Mental Model

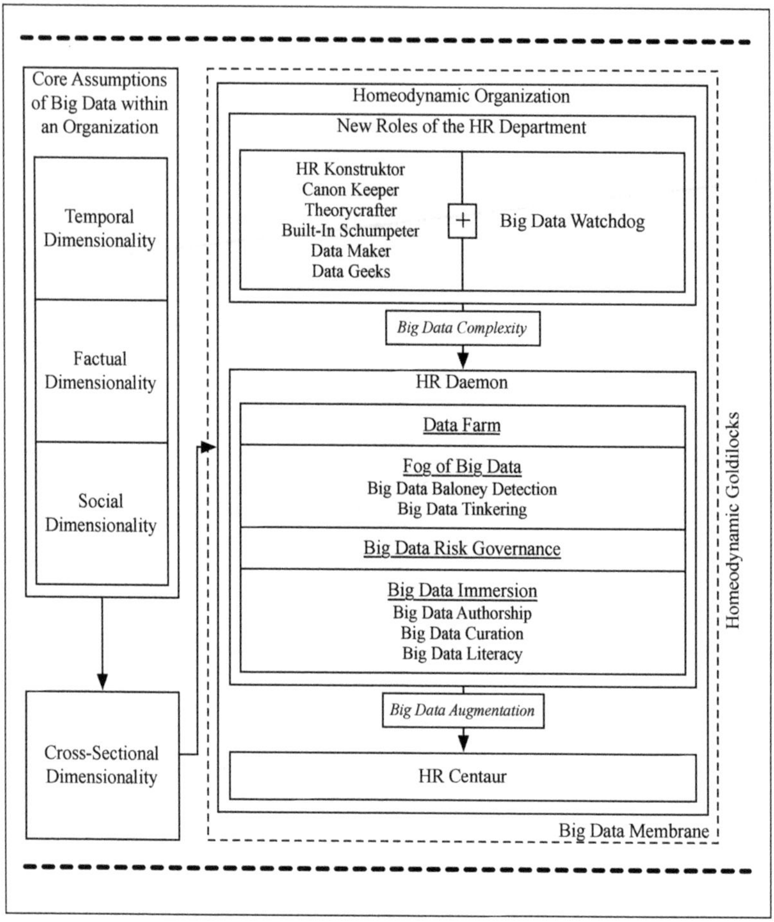

It is essential first to describe certain core assumptions regarding big data within organizations. Big data will have an impact on organizations and this impact will not be static but rather highly dynamic. These core assumptions are unique in all organizations, but influenced by temporal, factual, and social dimensionalities (following Stein 2000 on the basis of Kluckhohn & Strodtbeck 1961). Their arrangement, too, is unique for every organization and will be dynamic. These core assumptions will, therefore, merge into one distinct cross-sectional dimensionality. This unique dimensionality will act as the situational parameter on which an organization will depend, but will not be able to change in real-time; the organization will need to deal with it.

On the basis of the cross-sectional dimensionality, the general environment, and the influence of big data within organizations, I propose homeodynamic organization as a novel organizational type. It is derived from the homeodynamic concept introduced by Yates (1994) and, therefore, rooted in complex systems theory, but is expanded towards the need of dealing with big data. Consequently, any organization facing big data will transform into a homeodynamic organization and needs to react on this change. The driving force in dealing with big data will be the HR department. It is essential to highlight that big data in organizations will focus on the effect on the actors of said organizations, which means that employees are at the heart of my research.

The changes enforced by homeodynamic organization and, therefore, by big data, trigger a reorientation by the HR department. This reorientation will lead to new roles for the HR department. These new roles are oriented on the categorization of Ulrich et al. (2013), however, adapted to the unique settings within a homeodynamic organization. Therefore, I present six unique roles (HR konstruktor, canon keeper, theorycrafter, built-in Schumpeter, data maker, and data geek) and one cross-sectional role being the big data watchdog. All these roles are tackling certain aspects of the homeodynamic organization as well as the cross-sectional dimensionality introduced by big data.

However, this is just a response of the HR department to these fundamental changes, but as introduced in chapter 2.3, big data will increase the complexity within the organization. Reacting and changing the roles will not be sufficient, consequently, the HR department will create new structures within the organization. These new structures will mostly work in the background or the 'backend'. This construct that deals with big data and is created as well as implemented by the HR department will be called HR daemon.

The HR daemon comprises of a data farm that generates, cultivates, and harvests big data for organizations. The concept of a fog of big data is concerned with the problem that big data are not always precise and the challenges this impreciseness entails. It consists of a big data baloney detection which discovers faulty big data and big data tinkering which creates the possibility of exploring and searching for big data. Following this, the big data risk governance will be able to evaluate the risk of big data and combine it with the general risk, thus enabling the HR department to obtain a better sense of the potential risks and empowering senior management to make better decisions. The next component of the HR daemon is big data immersion dealing with certain aspects that are essential for handling big data and, consequently, required for any homeodynamic organization. It consists of big data authorship, big data curation, and big data literacy. Big data authorship tackles the question of data copyright and data privacy, thereby creating a solution that may work for organizations. Big data curation needs to keep big data in order and organized in a certain way, so that organizations do not drown in data. Finally, big data literacy refers to the HR department training employees in the adequate use of big data.

Big data are bound to change the organization as well as the role of the HR department extensively. But big data and homeodynamic organization depend on the proactive usage by all people within the organization. As depicted in chapter 2.4, big data are nothing that will be helpful if they stay in the backend; consequently, big data augmentation is a proactive goal of the HR department to increase the usefulness of big data within the organization. The homeodynamic organization requires a 'frontend' implementation that deals with interaction interface between people and system, in this case the HR centaur.

The HR centaur will enable employees to utilize big data for their purposes and increase the effect of big data on the organization as a whole. It is a way to make big data available and usable for everybody within the organization and, by that, to transform big data into a resource of pro-activity for homeodynamic organization, rather than the organization just reacting on big data. A big data membrane constrains the border of organizations. Big data are everywhere and there are no boundaries to them. The goal is to achieve a way of protecting certain parts of the big data and keeping them secure. Other parts of big data can be shared freely and openly.

Finally, homeodynamic Goldilocks will emphasize that a data-augmented homeodynamic organization will only perform well within a certain range and will be stable only if certain criteria are upheld. Big data will help achieve this goal but will likewise impede the process depending on the core assumptions made concerning big data in the beginning. For that reason, Goldilocks will be different for every organization.

3.2 Methodology

From a theoretical perspective, big data is still a relatively novel phenomenon. Big data have, however, a great impact on today's society, organizations, and individuals. Big data are currently lacking a concise theoretical foundation. Many researchers limit their view on big data to the perspective of a certain academic field. The prime goal of theory in general is to describe and explain (Whetten 1989), but big data challenge researchers due to their vastness. Researchers are unable to fully grasp big data; there will always be certain blind spots in any theoretical conceptualization.

The foundation of understanding big data is generated in the use of data. But data are not theories and will not automatically lead to theories (Sutton & Staw 1995). It is also evident that big data will not ever be understood entirely and that big data are too big for one grand theory alone. Any theory will always be an approximation (Weick 1995), so does one about big data. Big data and the concept of the homeodynamic organization both have complex and dynamic definitions, and any theory will be a lengthy interim struggle (Runkel & Runkel 1984). As Weick (1995) explains, there are few fully-fledged theories, and, therefore, big data cannot be made tangible by any comprehensive theory. It may be more fitting to 'theorize' big data and by that understanding big data as a more dynamic phenomenon. Weick describes theorizing as follows:

> "The process of theorizing consists of activities like abstracting, generalizing, relating, selecting, explaining, synthesizing, and idealizing. Those emergent products summarize progress, give direction, and serve as placemarkers. They have vestiges of theory but are not themselves theories. [...] The key lies in the context – what came before, what comes next?" (Weick 1995: 389).

If big data are all about data, it may seem obvious to consider the grounded theory (Glaser & Strauss 1967) and analyze data in order to create theories rooted in a positivistic view (Martin & Turner 1986). This may be especially fitting, as there is no theoretical framework available, since grounded theory does not depend on a theoretical framework (Allan 2003). It remains debatable, however, whether grounded theory leads to a theory or even contributes to theorizing (Suddaby 2006, Thomas & James 2006).

Another way of theorizing big data lies in a thought experiment or an experiment-in-imagination (Hempel 1965) that would anticipate the impact of big data on the basis of certain general rules and derive the outcome by means of deductive inference. In the context of big data, however, deduction may not be sufficient. Although the premise is to derive conclusions from the general to the specific (Samuels 2000), the experiment calls for the question: what is 'the general' in big data? Obviously that would be n = all, but that is not achievable (Junqué de Fortuny 2013, Ekbia et al. 2014, Forsyth & Boucher 2015). The basis of the literature is also highly dynamic (Thompson 1956) which is especially true for big data. Consequently, deduction in the case of big data would take place from the bigger specific to the smaller specific. Induction may, therefore, be more suitable, as it moves from special observations to general ones (Samuels 2000). That, however, sounds relatively similar to the social-constructivism or the proposed data-constructivism. A third form is abduction. The term was coined by Peirce (1958) and Hanson (1958). Gregory and Muntermann describe abduction as the method of "creating a theory [...] based both on real-world observations that are inductively observed as well as theoretical viewpoints, premises, and conceptual patterns that are deductively inferred" (2011: 8). The term gained popularity in the field of artificial intelligence (Bylander et al. 1991). Abduction is, in a sense, a way of combining induction and deduction. That, however, would be an oversimplification (Mayer & Pirri 1996). Induction and abduction highlight the data and deduction and abduction focus on knowledge creation (Shepherd & Sutcliffe 2011). Induction and deduction, however, are not sufficient to theorize big data. Abduction reveals the potential of bridging both elements.

On the premise of bridging induction and deduction, Shepherd and Sutcliffe (2011) developed the inductive top-down theorizing approach in order to establish a method of deriving new organizational theories. The goal is to connect induction, deduction, and abduction in a coherent approach. However, it may be debatable whether the authors are highlighting induction rather than abduction, especially as they state that the approach is "consistent with abduction" (Shepherd & Sutcliffe 2011: 361). Consequently, the name inductive top-down theorizing reveals a link to induction and deduction (through top-down) but no reference to abduction, although theorizing is concretized as "abductive theorizing" (Shepherd & Sutcliffe 2011: 371). The authors' intention was to incorporate earlier literature as well as data and new literature to build a new theory. Contrary to a solely deductive approach, the data and new literature "speak to the theorist (through the formation of gists) to focus attention so as to detect tensions, conflicts, or contradictions" (2011: 362). They also follow the general understanding that theorizing is an iterative process (Thompson 1956) and theories merely milestones. Consequently, the theorizing process becomes more critical, and as Weick argues: "We cannot improve the theorizing process until we describe it more explicitly, operate it more self-consciously, and decouple it from validation more deliberately" (1989: 516). The model of such a theorizing process is shown *explicitly* in Figure 7.

Figure 7: Inductive Top-Down Theorizing (Shepherd & Sutcliffe 2011: 366)

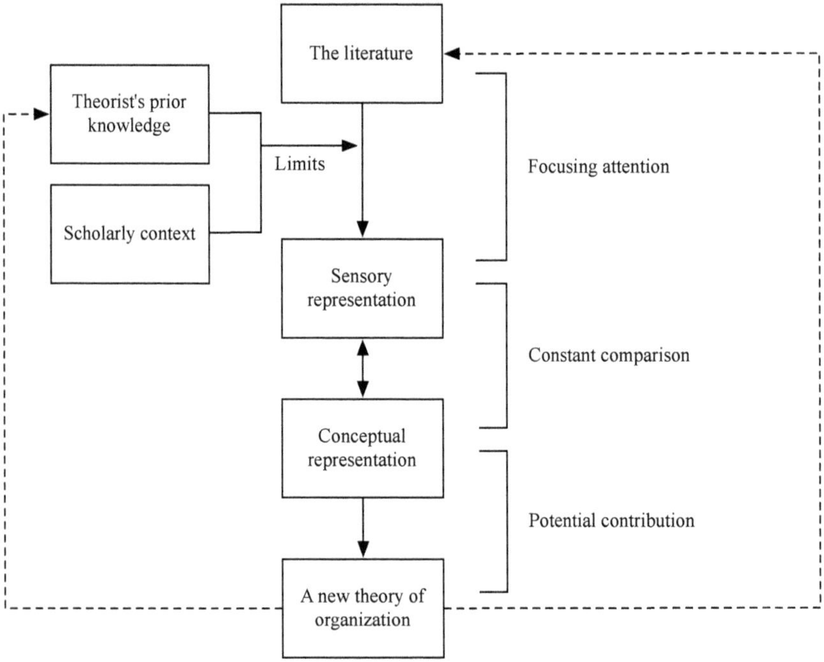

For Shepherd and Sutcliffe (2011), academic literature is the basis of research. It underlies constant change, however, and can consist of papers, books, presentations, working papers, and so on. Such a body of literature is massive and, therefore, research focuses a researcher's attention so that it be influenced by both a theorist's prior knowledge and the scholarly context. From a self-reflecting perspective, I tried to keep the literature I used as extensive as possible, especially as I have a background in organizational behavior, HRM, and information systems. I also made the acquisition of literature ongoing and took literature notes (Eisenhardt, 1989). As Shepherd and Sutcliffe (2011) describe, the scholarly context represents another influencing factor to a researcher. It is for that reason that working at a German university in the field of HRM and organizational behavior also had some influence on my theorizing process.

The focus of attention gravitates around the influence of big data within organizations and the influence on the actors within organizations. Technological elements of big data are reduced to social influences, and are not described in detail. Sensory representation is focused on humans and big data within organizations. In order to derive a new theory from this sensory representation, a step towards conceptual representation is required:

> "This *conceptual representation* refers to *general abstract statements of relationships between constructs – incorporating explanations of 'how' and 'why,' boundary conditions of values, and assumptions of time and space – that allow for a more coherent resolution of the theorist's sensory representation*" (Shepherd & Sutcliffe 2011: 366–367).

Both representations are compared to each other constantly in order to achieve a coherent picture. The authors claim that through the use of thought experiments and metaphorical reasoning, a convergence between both representations is possible. Thought experiments are similar to experiments-in-imagination (Hempel 1965). I will apply several thought experiments to existing examples of big data use and compare them to the general concept I have derived from the literature. Due to the vast disciplinary variety of sources of literature, metaphorical reasoning (Tourangeau & Sternberg 1982) will likewise become necessary to converge sensory and conceptual representation. Especially in the description of certain behaviors of big data, several metaphors as well as exemplary cases are utilized to describe big data more precisely. Big data are hard to grasp, and probably even more so is the way in which big data create reality. Metaphors are needed to describe the phenomena in more detail.

As a result, the thesis derives a new theory of organization through the inclusion of big data which makes it a potential contribution to the theoretical discussion on the effect of big data on society. Shepherd and Sutcliffe (2011) described four attributes that make a theory strong. The first one is its broadness and that it goes beyond one disciplinary field (Kilduff 2006). The second one is its simplicity and that it depends on few assumptions only. The third is the theory's concern with interconnections and interrelatedness. The fourth is that a theory has only few different explanations. A thesis may reach a certain outcome, but more importantly becomes a "stimulus for new theorizing" (Shepherd & Sutcliffe 2011: 374).

4. Analytical Implementation

4.1 Core Assumptions of Big Data within Organizations

Big data will influence the organization and will have a strong impact on any organization. In order to understand the nature of big data within organizations, it is essential to relate them with several core assumptions. While, as formerly depicted, the interaction between big data and humans is highly complicated, big data are a social phenomenon. Actor network theory defines technology as yet another actor and, therefore, frames big data as an actor within the social network of an organization. Stein (2000) postulates that, consequently, all members interacting in an organization are not only influenced by a social dimensionality, but also a temporal dimensionality and a factual dimensionality. That, as the author states, co-aligns with the structuration theory as postulated by Giddens (1984). Giddens (1979) stated that time-space relations are increasingly important for understanding social interactions. Gross understands Giddens as follows: "He argues that all social systems must be understood as stretching over time and space, or better, 'embedded' in time and space" (1982: 83). This, to a certain extent, negates the former statement that actor network theory and structuration theory do not fit well together since, from the perspective of time and space, they are not contradictory. Law (1992) reports the ordering potential of time (durability) and space (mobility) within systems, thereby, as well as highlighting the importance of time and space within social systems.

In addition to the relevance of time and space, Kluckhohn and Strodtbeck (1961) discuss the dimensionalities of the senses, and classify them into temporal dimensionality, factual dimensionality, and social dimensionality. Space is absorbed within the factual dimension and is expanded. Stein (2000) uses these dimensionalities as his core assumptions for the developmental analysis of organizations. The temporal dimensionality deals with time and consists of an assumption about the *direction of time* and *velocity*. The factual dimensionality goes beyond the concept of *space* and also includes assumptions about *reality* and *risk*. The social dimensionality involves the way in which organizations assume their *identity*, *action*, and *trust*. In enhancements of his model, Stein (2000) proposes that any of these core assumptions can be described as polarities, and that organizations range along the spectrum of those polarities in the sense of an overall profile. I use these core assumptions to describe the polarities of the views that organizations hold of big data and the way they are being handled. Organizations also consciously position themselves within the spectrum of any of these core assumptions from which they derive strategies and operational structures. Table 11 presents the related polarities of big data within organizations.

Table 11: Polarities of Big Data in Organizations on the Basis of the Core Assumptions

Core Dimensionality	Core Assumption	Polarities	
Temporal Dimensionality	Direction of Time	Data Linearity	Data Monadology
	Velocity	Data Rigor	Data Swiftness
Factual Dimensionality	Space	Data Island	Data Assemblage
	Reality	Social Constructivism	Data Constructivism
	Risk	Data Risk Avoiding	Data Risk Seeking
Social Dimensionality	Identity	Social Shadow	Data Shadow
	Action	Self-Determined	Data-Determined
	Trust	Data Reliance	Data Bias

These core assumptions will be used as the starting point for understanding big data within organizations, as well as to refine the question of what an organization faces concerning big data. The polarities will describe the effects or outcomes of big data within an organization and not the implicit notion of integration design (e.g. Stein 2014). Organizations can be steered into a certain direction on the spectrum, with the goal to steer them in a certain direction, ideally to support the homeodynamic stability and agility of the organization.

4.1.1 Temporal Dimensionality

Time is important, especially in the context of organizations in which "time is money" (Loft 1995: 127). Big data are sending mixed signals, though. On the one hand, big data are available in an instant, on the other hand, big data are so ubiquitous that organizations are overwhelmed and need time to cope with the abundance. Dealings with big data are linked to the temporal dimensionality, and organizations need to consider the possibilities of integrating them. Big data are susceptible to changes in the temporal dimensionality and will influence future big data, as big data constantly generate new big data over time.

Data linearity or data monadology. Big data can also be seen as a temporal construct in itself. Any data within big data are linked to a timestamp, be it the time they were collected or the time of the collected incident. An example may be a historical book by a contemporary witness and a book by a researcher. The first book has a timestamp of the respective time period when something originally happened and the latter one has a timestamp of more recent years. Both books refer to the same event and will (hopefully) include similar data among the information they convey. Nevertheless, both have vastly different timestamps. Such differences raise certain obstacles concerning big data. One way of coping with the direction of time is to see big data as a *linear* construct. Data in history are added up in a linear way and new data are constantly added to the tail of a linear stream of data. This view would decrease the complexity of big data drastically, as big data could be transformed into a timeline. In the context of organizations this seems particularly plausible due to the obsolescence of information (Argawal et al. 2005) and the half-life of knowledge (Machlup 1962). Organizations can focus on the most current data. *Data linearity* is consequently one-directional, and, therefore, obsolete information is *unlearned* or, more precisely, buried beneath new and momentous information. In times of big data and the potential danger of data avalanches (Miller 2010) under the assumption of linearity, this is a plausible concept.

An alternative perspective is seeing big data not as a linear but as a *non-linear* construct. Focusing on the non-linear perspective is similarly interesting for organizations. Big data that are relevant or related to an organization are essential to said organization. Although the flap of a butterfly's wing on the other side of the world may have an influence on an organization, the chances of occurrence are so infinitesimal that investing time and resources in order to prevent it is not efficient. Big data within organizations are, therefore, always merely just a portion of all big data and the organization itself is the one to select the relevant portion. This conceptual view resembles the idea of monadology, or the theory of monads (Tarde 1893/2012) which Latour et al. (2012: 598) describe as follows: "A monad is not part of a whole, but a point of view on all the other entities taken severally and not as a totality." Although this idea conflicts with the observer problem with big data, seeing an organization as a monad is helpful in understanding a non-linear perspective on big data. Latour (2002) argues that it is essential to move beyond a micro/macro categorization, and I propose that big data can be seen in a similar way. There is no obsolescence due to time, but there is obsolescence due to the monadological and non-linear connection. Certain elements of big data are irrelevant for certain organizations while other elements are relevant. From this perspective, organizations have a "highly specific point of view" (Latour et al. 2012: 598) on big data, and this view is decoupled from the linearity of time. Tarde (1893/2012) treats time following the argument of Leibniz, as not absolute but relative and rootless to non-linear connections. In the following example, although Giddens does not mention the connection to Tarde, time is not of relevance, the non-linear connection of the words unveils the meaning and story behind the example (Giddens 1984: 302).

> Private property : money : capital : labor contract : industrial authority
> Private property : money : educational advantage : occupational position

This example reveals the translation, or transformation of private property into something different. On the basis of the monad, however, private property is embedded into different contexts. Both transform their private property (although not exactly as defined by Giddens) into money. The first monad uses the money to gather capital, contract new labor, and achieve industrial authority, the other monad uses the money to gain an educational advantage that leads to a better occupational position. Although the time is unknown, the non-linear progression reveals two different stories. The first monad is probably an employer while the second one appears to be an employee which shows that a *data monadology* is non-linear. Organizations select and utilize relevant big data. Following this logic, the sequence of combining big data will become more important, as will navigation through big data. Especially under the assumption that big data generate new big data, any monadological step will generate big data that depends on the non-linear perspective of the monad/organization. Presuming linearity or monadology will, therefore, shape big data within organizations.

Data rigor or data swiftness. The next core assumption regards velocity. One major attribute of big data is that they can be analyzed quickly, data streams can potentially be analyzed in real-time (Barlow 2013). Analyzing big data in such a way comes with a certain tradeoff. Such analyses can be described as *data swiftness*, and although the results are nearly instantaneously available, those analyses may not be very precise. They are often designed without any hypotheses and favor correlations over causation. The use of big data in this particular way, therefore, is susceptible to errors. At the other end of the pole, there is *data rigor* use. Such precise and thoughtful use comes with a hefty toll on velocity. It takes time to analyze big data in that manner, although such an analysis is less prone to errors and gives more detailed insights into organizations. Such results are also evaluated and can explain causation within the data set.

Table 12: Big Data Tradeoff Concerning Velocity

		Data Swiftness	
		Low	**High**
Data Rigor	Low	Undesirable	High Data Swiftness
	High	High Data Rigor	(Currently) Impossible

As shown in Table 12, organizations face a decision concerning their direction of big data analyses. They can choose between high data swiftness and high data rigor. Other combinations are either impossible (high data swiftness and high data rigor) or undesirable (low data swiftness and low data rigor). Marketing methods applying the shotgun principle, being high in data swiftness, are promising; they may for example lead to an increase in sales (Mayer-Schönberger & Cukier 2013). Analyzing data from experiments like CERN, on the other hand, needs to be rigorous and will take time (Wright 2014). Organizations freely choose the way in which big data are used and will deal with the consequences. Organizations evaluate the costs they generate through velocity and the costs of the potential errors that may result from excessively rapid big data analysis.

The time perspective reveals that organizations can tackle big data in a variety of ways, however, decisions made by the organization influence the integration of big data into organizations. They can assume that big data will be linear or that it follows the logic of monadology, and decide between high data swiftness and high data rigor, but both influence the precision of the results of any big data analysis. Although such polarities apparently propose an either/or decision, organizations have the ability to apply both polarities. They can decide this before every big data analysis or even utilize both polarities, starting with high data swiftness and finishing with high data rigor. While possible, the likelihood of such an approach being taken in reality is debatable considering the increase in cost.

4.1.2 Factual Dimensionality

For this dimensionality, Luhmann (1991) uses the German term *sachlich*, which translates to 'objective' or 'factual'. Luhmann was predominantly concerned with the discourse about reality, subjectivism, and social reflexivity. Using the translation 'objective', therefore seems inadequate – especially concerning the problematic subjectivity of big data and their appearance as objective. The factual dimensionality deals with the tangible influence of big data on organizations and the ways in which an organization can use big data to transform itself. Big data provide a massive amount of information capable of influencing the factual dimensionality within organizations. Space, reality, and risk are affected by big data, but the direction of said influence depends on the underlying assumptions that organizations make concerning the comprehension of big data.

Data island or data assemblage. Stein (2000) points out the potential of technology to bridge space. In recent years the spatial distance has decreased significantly (McCann 2008). Space in the sense of spatial distance is no longer adequate for understanding the obstacles or polarities that space entails in organizations. Big data in particular contribute to the lack of spatial distance. In recent years, society has unlocked a new form of space that is parallel to its classical form. The internet has contributed to the concept of virtual space and big data help making this virtual space ubiquitous (Giard & Guitton 2016). There is a complete virtual dimension parallel to normal, or real space. Virtual and real space are not separate from each other and current developments indicate that both spaces are moving into alignment with each other (Bimber & Raskar 2005). Both worlds are permeable and people seamlessly jump from one world to the other. Navigation within real space, for example, is often accomplished using tools from virtual space. People often no longer consult paper maps; they use Google Maps. This represents an evolutionary development of society that, through augmented reality, gradually merges the two worlds (Azuma 1997). Big data are a main driver of this change (Swan 2013). Organizations, therefore, reconsider their own design as well as they picture the membrane between real space and virtual space.

It may be possible for organizations to regulate big data and strictly control their use. Using a metaphor from a spatial perspective, big data can be understood as on an island. Only a small number of people within each organization has a boat to steer to this *data island.* Big data will be spatially far away from organizations and interaction will be limited and closely monitored. Organizations are establishing artificial distance in a figurative sense, which enables them to steer the internal effects of big data. Big data are, once more, placed inside of an iron cage while their use is regulated with an iron hand. Organizations could follow the idea if they were to assume that big data were acting as something uncontrollable and uncertain. Big data can change structures and change organizations and those developments may turn out to be structural shackles (Scholz 2015a) which may bring about the tendency to prohibit such a rampant use of big data and put big data on a data island. Such an assumption will probably tie up a big portion of an organization's resources, however, and may, therefore, be more efficient, assuming that big data help converge real space and virtual space together. At the very least, this is preferable to isolating big data within organizations.

Both spaces can be seen as permeable, and the organization as a real space will be open for interaction with virtual space that is big data. Such a concept resembles the concept of habitus (Bourdieu 1977) because "habituses are permeable and responsive to what is going on around them" (Reay 2004: 434). Bourdieu (1977) claims that habitus is both *opus operatum* (the result of practice) and *modus operandi* (the mode of practices) which is applicable to both the organizational habitus and the relationship between big data and the organization. Habitus appears as a fitting theory through which to understand the relationship between real space and virtual space, but Morrison (2005) as well as Reay (2004) note that there is a latent determinism, a focus on continuity, a neglect of change (Shilling 2004), and a strong emphasis on structures (Bourdieu 1986). A different notion of permeability between real space and virtual space links to technology as well. If both spaces are seen as equal (in analogy to Bryant, L. R. 2011), they all are actors and contribute to organizing the organizational network. Parker (1998) calls it 'cyberorganization' and combines real space and virtual space as a new form of organization. In recent years, the term 'assemblage' has gained popularity when describing this interplay between both spaces (Taylor 2009). Kitchin (2014a: 24) defines data assemblage as the "composition of many apparatuses and elements that are thoroughly entwined, and develop and mutate over time", but he sees the concept as predominantly connected to the production of data (Kitchin & Lauriault 2014). In this context, on the other hand, data assemblage is defined as an interrelationship (Giddings 2006), and a dynamic process (Taylor 2009). It is no longer possible to differentiate between real space and virtual space. Organizations permeate big data and big data permeates every organization. Generally speaking, *data assemblage* sounds a more realistic approach to big data within organizations. However, supposing such an overlapping of real space and virtual space makes it difficult for any organization to deal with big data altogether. It may even be assumed that resistance is futile (Russom 2013). To recapitulate the assumption about space, organizations decide on a spectrum between doing nothing against big data and letting it flow through the organization, or restricting and limiting the use of big data completely. Both poles are probably too extreme. The way organizations first want to understand the relationship between organization and big data, however, will be a strategic decision.

Social constructivism or data constructivism. The next core assumption is about reality, an issue that is picked up numerous times in the course of this thesis. The conclusion of this discussion is that reality is constructed and, even though big data are vast and ubiquitous, they are incapable of representing reality as an objective truth. Big data are no Laplace daemon and will probably never be capable of being omniscient (Scholz 2015a), which results in the idea that big data create reality as well. Reality is created either on the basis of *social constructivism* or *data constructivism*. Translating this into organizational interaction with big data means that organizations will either influence big data or big data will deliver relevant information and insight. Conversely, data constructivism is the idea that big data influence organizations in such a way that the surrounding reality is shaped by big data. This development has recently become observable in the discussion about data-driven decisions (McAfee & Brynjolfsson 2012). Although for legal reasons, the decision needs to ultimately be made by a person, this person decides on the basis of the information provided by big data. Decisions are shaped by the reality constructed through big data.

Both assumptions have an impact on how organizations will work in the future. Assuming the social constructivist view, the impact will be similar to the beliefs of the neo-luddites, and following the data constructivist path would be more like the ideals of the anti-guessworker. However, both poles will inherently distort the reality of organizations, in a certain way acting as a reality distortion field (Levy 2000), and this subjective reality will be reinforced over time. Both views are, therefore, highly susceptible to objective subjectivism (Gadamer 1992). Big data within organizations force them to choose a certain path and deal with the consequences. Contrary to other assumptions, constructing reality is similar to a path-dependence (Sydow et al. 2009) and a lock-in (David 1985). A de-lock-in can facilitate changing the path, but achieving such change and negating the reality distortion field of organizations takes time.

Data risk avoiding or data risk seeking. The final core assumption concerning the factual dimensionality regards risk. Dealing with risks is essential for the survivability of any organization. Generally speaking, people and therefore also organizations can be categorized according to their risk behavior: as either avoiding risks, being neutral towards risks, or seeking risks (Kahneman & Lovallo 1993). Risk avoiding and risk seeking represent the polarities of this spectrum. Although these polarities are nothing new and big data will not add new facets to these characteristics of people (e.g. Tallon 2013), they will have an amplifying effect on both polarities. Big data can help a risk avoiding organization become extremely risk avoiding, especially since many risks can be discovered by means of big data. By calculating every potential risk in the evaluation, it is possible to avoid them altogether. Conversely, a risk seeker will have the same information, but will come to a different conclusion, likely taking the risk regardless of the information supplied by big data.

Within organizations, there is a variety of different types of risk behavior, but depending on the general attitudes to risk, big data can be shaped accordingly and even falsified accordingly. Risks may be increased or decreased by big data within organizations. Big data are, therefore, a new risk factor and these assumptions of risk and the risks which result from them need to be addressed by organizations – no matter the polarity in which they lie. I propose the concept of *big data risk governance*, which will be described in the course of this thesis.

The factual dimensionality specifies that big data will lead to a new understanding of space and, consequently, have a strong impact on the idea of space within organizations. Big data will also challenge reality, not in the promise of objectifying reality within an organization, but by being a new source of constructivist direction within organizations. Finally, big data amplify the potential for risks in organizations. Although there is an underlying assumption of polarities, organizations will mostly find a position along the spectrum and not at the extremes. Nevertheless, the dimensionalities highlight the essential need for assuming a certain understanding of big data within organizations. Simply using big data without consideration will have long-term consequences that cannot easily be repaired.

4.1.3 Social Dimensionality

The final dimensionality tackles the relationship of people with each other and the difference between the *Me* and the others. It asks the question of consensus or dissenus and the underlying morality (Stein 2000). The social dimensionality is concerned with the relationship of the individual with other actors in their surrounding organization, and big data are amongst the actors with which the individual interacts. The individual estimates the role of big data within organizations and the effects of big data on organizations, as well as the role that big data will play with regards to the individual. Such assumptions will influence the function of big data enduringly for the future. Assuming that big data will change an individual's life for either better or worse will cause the individual to act differently, and will affect the individual's identity, actions, and ultimately the trust the individual has in big data, as well as the way other people and big data perceive the individual within organizations. Big data add a new perspective to the social facets of an organization, the way in which people interact with each other through big data, and, more importantly, how they are influenced by information from big data.

Social shadow or data shadow. Identity is an important part of an individual and reveals its uniqueness. It entails a sense of self-conception and the idea of a person being different to others. But there is a potential difference between the self-perception of an identity and the way in which a person is perceived in their social surroundings. This may be the result of social stereotyping or because a person is wittingly acting in an atypical way. Identity can be assumed to be comparable to a black box, only giving insights through interaction with external environments. Neither big data nor social interaction will give a precise description of an actual identity, never being more than shadow. While this shadow may be granular, it could also be a shadow play and be completely different to the actual identity.

Big data add a new form of shadow to the perception of identity. Haggerty and Ericson (2000) describe this new digital identity as 'data doubles'. On the basis of the Orwellian increase in surveillance that mimics a panopticon, they propose the idea that people are doubled within big data. This idea was picked up by Wolf (2010), Kitchin (2014a) and Scholz (2015a), and expanded into the concept of data shadows. Wolf claims that people cast data shadows wherever they go, and Kitchin describes those data shadows as "information about them generated by others" (2014a: 167). Scholz defines data shadows on the basis that big data are subjective and contextualized and "we are only seeing the shadow of reality (comparable to the allegory of the cave by Plato)" (2015a: 8). Even with big data, peoples' view is, therefore, limited to the shadow of the identity of others. In addition to that, the subjectivity of big data may distort the data shadow. Big data attempt to double or copy the original and exhibit in this way similarities to the concept of a simulacrum, defined as:

- "it is the reflection of a profound reality;
- it masks and denatures a profound reality;
- it masks the absence of a profound reality;
- it has no relation to any reality whatsoever: it is its own pure simulacrum" (Baudrillard 1994: 6).

Big data try to achieve a simulacrum of people by analyzing their digital footprints (Sellen et al. 2009) and data trails (Davis 2012), thus, using those breadcrumbs (Cumbley & Church 2013) to reflect the identity of that person. However, all four definitions of simulacrum are possible and make it difficult to achieve convergence between identity and data shadows. Interestingly, Baudrillard hints at the strong impact of the current explosion of data, which will lead to an implosion of meaning: "We live in a world where there is more and more information, and less and less meaning" (1994: 79). The author predominantly refers to this information exhaust in terms of the media, but he reasons that "information devours its own content. It devours communication and the social" (1994: 80) and, therefore, picking up McLuhan's (1967) formula, the medium is the message. Indeed, Baudrillard (1994) suggests that the media will create a simulacrum that simulates a hyper-reality. Hyper-reality is the sense of an inability to distinguish between reality and simulacrum (Tiffin & Nobuyoshi 2001). In the context of big data, identity is depicted, interpreted, and transformed on the basis of its data shadow, into a simulacrum that eventually creates a *hyperidentity*. People unwillingly but constantly contribute to their hyperidentity without having any control over it (Pasquale 2015). This hyperidentity can be a granular reflection of an actual identity but also has no relationship whatsoever. In some cases, for example, people are evaluated on the basis of their residential address and the behavior of other individuals in their area. If a person lives in a low-income area, gaining credit may become a challenge (Pasquale 2015) which goes to show that an individual's hyperidentity has no connection to their actual individual identities.

This form of shadow is created by big data and people have no grasp of the entirety of information contributing to their hyperidentity; however, individuals can wittingly influence their *social shadow*. As Goffman (1959) explains, people are capable of interacting with other people differently. He compares this to a theatre, where there is a difference between the identity of actors on stage and their back-stage identity. People play an act on stage and their identity is perceived mostly in regard to their acting. They put on masks or costumes and simply become different people. For that reason, within an organization that can be compared to a stage in a theater, people act in a certain way and thus create a stage identity. Such a stage identity is perceived by the audience at the risk of perceiving the actors in a slightly different way than the actors themselves (Watson 1982). The following statement describes the mismatch between individual identity and social identity quite clearly: "If one knows who one is (in a social sense), then one knows how to behave" (Thoits 1983: 175). People contribute to their stage identity while simultaneously having to juggle several stage identities at the same time (Scholz 2016a). In times of big data, playing an act in a certain stage reality will become increasingly difficult as everybody is constantly under surveillance and no longer capable of separating stages precisely. Hyperidentity and stage identity influence each other. A Facebook identity, for example, influences a professional identity and the chances of being recruited (Ramo et al. 2014). The interrelationship between both identities and the difference to the actual identities is summarized in Figure 8.

Figure 8: The Perception of Individual Identity on the Basis of Data Shadow and Social Shadow

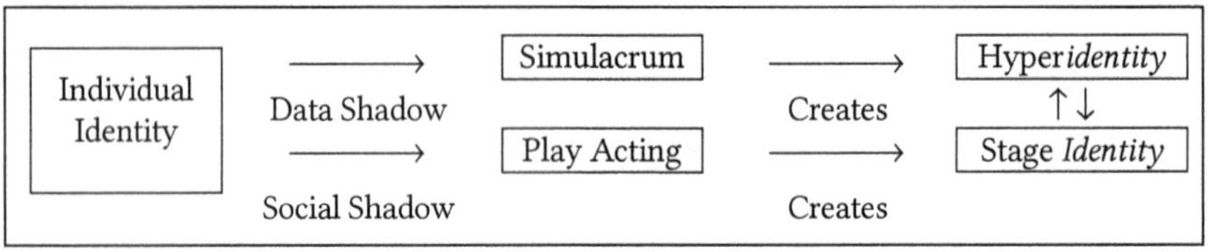

Self-determined or data-determined. The detection of a certain shadow within organizations is linked with the next core assumption of action. Because a data shadow is not much influenced by the individual or the organization, this implies a data-driven understanding of big data. If, however, the individual and organization assume a certain form of social shadow, they are capable of changing their perceived identity. Such an assumption will lead to a certain perspective on determination. The question is whether the individual and the organization are *self-determined* or *data-determined*. Self-determination in the general sense refers to the self-motivation or intrinsic motivation required in order to achieve certain goals (Ryan & Deci 2000). In that case, the individual at first has the motivation to achieve a certain goal, and afterwards uses big data as a tool to achieve said goal. Data-determination is the concept of externally motivating people to achieve a certain goal. There is much discussion about big data nudging (Thaler & Sunstein 2008, Yeung 2016) things in a certain direction, especially in the context of big data. Richards and King (2013: 44) describe this nudging as follows: "The power of big data is, thus, the power to use information to nudge, to persuade, to influence." Big data supply the individual and organization with enough information to influence their decisions which is already done by politics (Nickerson & Rogers 2014) and governments (Schroeder 2014).

Self-determination, therefore, can also mean that big data are nudged in a certain direction. If big data reveal the desirable goals, the organization is purely data-driven. If big data are considered as a tool to achieve certain goals, however, the organization is still self-determined. Both polarities have the ability to nudge the other. However, for any strategical decision within organizations, the dominant pole is as clear as who nudges whom.

Data reliance or data bias. Finally, there is the core assumption of trust. People may believe in big data and the correctness of big data, which leads to a certain form of *data reliance*. As some suggest (e.g. Anderson 2008), if we have enough data we will get results. Such a core assumption is possible and is used (Servick 2015), but there is also the assumption of a general inherent *data bias* (Scholz 2015a). Although as stated earlier, big data rely on a certain form of data bias, the assumption of such bias will lead to the belief that big data in general are incorrect and maybe are not even used at all. Similar to the views on big data within HRM, the two views overestimate both the objectivity and the subjectivity of big data. However, individuals in an organization will act on said assumptions and either be open or skeptical towards big data within the organization.

The social dimensionality reveals that those core assumptions will have an influence on the actual use of big data within organizations. They raise the discussion about whether decisions have become driven by big data or whether people are still capable of deciding on their own, especially in the context of identity and its perception. Do people perceive others through data or through social interaction, and how do those ways of perception differ? The social dimensionality highlights the concept of distortion through both big data and through social interaction. Assuming a certain perspective will have an influence on the other polarity and, thus, reveal the potential for nudging the other pole. Big data can nudge people in a certain direction, and people as well can likewise nudge big data in a certain direction. This development can lead to a vicious cycle. This is especially true given the difference between being self-driven or socially-driven, since being data-driven may imply that people and their decisions are controlled externally.

4.1.4 Cross-Sectional Dimensionality

Big data are adding new types of dimensionality to organizations on top of potential existing dimensionalities. Therefore, the complexity increases and this can be compared to the curse of dimensionality – a term coined by Bellman (1957). The curse explains the phenomenon that, on adding new dimensionalities of data into for example an algorithm, the algorithm will need increasingly more time to deal with these new dimensionalities. Nevertheless, adding new dimensionalities will make the results more precise. That is the same case for the core assumptions based on the presented dimensionalities. However, they are currently separated from each other and need to be connected. Therefore, some sort of cross-sectional dimensionality is required, despite the curse dimensionality that arises.

In summary, big data are not objective and organizations are no longer comparable, and any organization can take on different points of view regarding big data. All of which being potentially fitting, however, claiming such subjectivity and uniqueness of organizations and, furthermore, proposing that big data can increase and decrease variety in an organization, may render organizations more standardized or more singular. Converging towards a certain standardized solution will lead to a loss of competitive advantage and a reduction in the survivability of those standardized organizations. A competitive organization will, therefore, prefer to become more singular rather than more standardized.

Holzkämpfer (1995) discusses singularities within organizations and focuses on extraordinary incidents within organizations and exceptional structures of organizations. Big data can be described as following his vision and will lead to such structures assuming big data have a general tendency to become more unique in order to increase the survivability of organizations.

Singularities are seen differently in related literature. There is, for example, the concept of the technological singularity. A technological singularity is the point in time at which artificial intelligence has increased technological progress to an uncontrollable and unpredictable extent (Kurzweil 2006). Although big data contribute to the manifestation of this hypothetical point in time, this thesis is concerned with singularities within dynamical systems and, therefore, rooted in systems theory, cybernetics and ultimately complex systems theory (Holzkämpfer 1995). Holzkämpfer traces the term back to Poincaré (1881) and Maxwell (1882). The latter describes the influence of singularities as follows: "it is to be expected that in phenomena of higher complexity there will be a far greater number of singularities" (Maxwell 1882: 443), thus, strengthening the relevance of singularities in highly dynamic systems. Social systems in particular are complex and dynamic. For that reason, "a small error in the former will produce an enormous error in the latter" (Poincaré 1914: 68). Holzkämpfer (1995) proposes the idea that organizations are influenced by singularities. He defines the following features of singularities:

- "Instability
- System-relatedness
- Uniqueness
- Irreversibility
- Subjectivity
- Randomness
- Complexity
- Reciprocity" (Holzkämpfer 1995: 91).

Even without describing those features, it makes sense to view an organization affected by singularities. Big data will contribute to this perspective and, furthermore, can be seen as a set of singularities as well, a phenomenon I propose calling the *big data singularity*. Describing an organization affected by big data, and regarding that big data as singularities, fortifies the claim that any organization deals with big data in a unique way. Instability is used in the sense of singularities where small things can have large consequences, and such ripple effects become obvious in big data and organizations. Due to the amount of big data, numerous minor things may have an influence on an organization. System-relatedness emphasizes that certain data are relevant to certain organizations. Not all big data are relevant to all organizations, only big data compatible with the organization's context will have an influence. Big data are contextualized and cannot be simply applied to a context they are not intended for. Organizations are unique; every organization is, in its way, truly singular. The same applies to big data; a body of data is singular as well. On the basis of the increasing granularity of big data (Kucklick 2014), every data point is unique. Data can vary in information, but the way data are predominantly collected makes it nonrecurring. Data can be collected with various tools, at distinct times, from a certain perspective, and with different intentions. While data may appear similar, metadata are always different and render every data point unique.

Any change within an organization is irreversible. More precisely, any change will create a new organization, and by trying to reverse the change, organizations will remember the recovery process. An organization is, therefore, not reversed to its former state but shifts to a new state merely resembling the old stage. The same principle applies to big data as well. Any change in big data will irreversibly transform them in a certain way. Such changes cannot be retracted, and repairing them is extremely difficult. Big data float everywhere and changes will spread through them. A certain data point can be changed in a distinct location, but whether or not the information it contains will be changed elsewhere remains unknown. Pasquale (2015) describes this phenomenon using the example of credit scores which can be changed, but not throughout all data that constitute an individual's credit score. Big data are also subjective, which imposes constraints on both big data and organizations.

The next aspect that contributes to the singularity of big data is randomness; generally speaking, those aspects that can be observed in an organization in terms of cause and effect are random. Big data have difficulties identifying causalities and often uncover correlations. They are also far from n=all, and no organization will ever have access to the totality of big data. Any selection of big data is to a certain extent random in itself. Both organizations and big data are highly complex and big data are complex. They also have a reciprocal relationship, which, while adding to their complexity, also contributes to the creation of new singularities through interaction between organizations and big data.

Creating an organization that is dynamic or even homeodynamic is plausible. Adding the concept of singularity derived from cross-sectional dimensionality to the mix allows an organization to understand and grasp big data in a more appropriate way. Since big data are comparable to singularities, every interaction with big data is seen as something novel and distinct. Organizations will also move more in the direction of dealing with singularities from a structural perspective. Contextualizing an organization, therefore, needs to be more dynamic and relational. Dourish (2004: 22) developed a different view on contextualization: "contextuality is a relational property, the scope of contextual features is defined dynamically, [...] context arises from activity". On the basis of this idea, Scholz (2013b) developed the concept of relational contextualization. The use of 'relational' in terms of contextualization may seem tautological at first glance. It does, however, underline the point that relational interaction with the context is an increasingly important factor for understanding organizations. Big data in particular become more useful if the relationships in which they are generated are transparent. Relational contextualization is essential for understanding organizational singularity, big data singularity, and, consequently, the reciprocity of the two.

4.2 Homeodynamic Organization

4.2.1 Characterizing Homeodynamic Organization

Big data and organizations constantly influence each other and are embedded into a dynamic and turbulent environment. On the basis of such an extensive relationship and that big data have a new influence on the organization, it makes sense that organizations will change. The core assumptions and the presented polarities reveal that organizations will have some freedom to react and create a unique response towards big data. Consequently, big data will trigger a transformation, but organizations will respond with a dynamic approach. The situation of the organizations can be compared to the causal texture turbulent field (Emery & Trist 1965), in which processes are dynamic and an organization is strongly interconnected with the field. This field is subjugated to linearity and non-linearity at the same time, and subsequently to order and chaos. Organizations, thus, try to achieve a certain form of homeodynamic balance in order to increase their survivability. The goal is to achieve a dynamic stability and a temporary equilibrium within the general imbalance (Luhmann 1991). In this context, however, stability does not refer to the steady state of homeostasis, but the ability to keep organizations alive. More fitting is the analogy with nautical terms: stability means keeping the boat steady or staying on a steady course. Although the ship is influenced by the environment and depends on its own integrity, the helmsman's task is to take account for all these factors and keep the ship stable.

Modern organizations are comparable to such ships, as organizations will be kept on track in order to stay profitable and, consequently, survive in today's stormy environment. Successful organizations will, therefore, act more like homeodynamic organizations that seek a homeodynamic balance. The concept of 'homeodynamics' as used in the course of this thesis was introduced by the following definition:

"Homeodynamics [involve] rate-oriented homeodynamic stability, not very far from equilibrium, fluctuating and oscillating or close to 1/f noise informationally, not fixed program-driven systems with easy generation of new activity patterns" (Trzebski 1994: 111).

Yates (1994) presents several errors concerning living systems, from which he derives the homeodynamic concept. An organization can also be seen as a living system (Kast & Rosenzweig 1972), and, following the argumentation of Yates, homeodynamics can be applied to modern organizations. One aspect of homeostasis, as criticized by Yates, is the idea that such systems are state-determined in the sense that such states influence the rate of the system. However, he makes a distinction between the two aspects homeostasis and homeodynamics in reference to non-linearity. More relevant for living systems is the rate or the velocity at which they are influenced by and influence other living systems. They have a tendency to be not very far from some form of equilibrium, and oscillate around both the equilibrium and the noise (or disorder). Living systems may be constantly changing. Yates also considers the program-driven idea of living systems. They are not predestined by their DNA and are capable of change which renders them more execution-driven. New activity patterns are constantly generated to cope with new challenges. Finally, Yates presumes that: "Systems are dynamically stable, meaning that they are able to sustain their trajectories in their basins of attraction even when coupled with dynamically rich inputs that can overwhelm structural stability" (1994: 70). In saying this, Yates moves beyond the idea that stability is linked to structure, thus agreeing with Farjoun (2010) and his concept of change as stability. Interestingly, Yates claims that homeodynamics can serve as a meta-theory (1994) for understanding living systems, but he also demonstrates that a new complexity theory needs to emerge that "must ultimately displace cybernetics, general systems theory, artificial intelligence, dissipative structure theory, information theory, and control theory from their fashionable apotheoses" (Yates 1994: 71).

When translating these ideas and concepts into a homeodynamic organization, there are several aspects that are important for an organization. First of all, any organization is changing constantly and will, over time, evolve into a new dynamic system: "homeodynamics refers to the continuous transformation of one dynamical system into another through instabilities" (Lloyd et al. 2001: 136). But organizations are subservient to attractors and will flow between them. Lloyd et al. (2001) claim that an organization tends to behave homeodynamically if there are large attractors. Big data will serve as a large attractor and, therefore, foster the tendency of organizations to become homeodynamic. An organization will be able to self-reconfigure and be dynamic enough to achieve reconfiguration in a quick and precise way.

In the realm of organizational theory, this description fits with the idea of dynamic capabilities. Teece et al. (1997) describe dynamic capabilities as the ability of an organization to use and reconfigure competences in order to deal with environmental changes. According to the dynamic capability approach, organizational competencies are achieved by smartly combining organizational resources.

One crucial goal of dynamic capabilities is that the resource configuration of organizations does not fossilize, but remains dynamic and flexible. Dynamic capabilities depend on an ongoing monitoring of resource allocation and ongoing competence-specific resource reconfiguration (Eisenhardt & Martin 2000). By being able to add new resources and release obsolete ones (Wang & Ahmed 2007), organizations stay flexible and dynamic in generating new competitive advantages (Sanchez et al. 1996). They therefore contribute to rate-oriented homeodynamic stability.

It is essential for organizations to act close to the equilibrium and close to the noise. As covered by complexity systems theory, organizations simultaneously operate at the edge of order and the edge of chaos at the same time. Big data act as a novel and potent source of disturbance. Organizations utilize this perturbation in order to gravitate around both edges. This idea exhibits consequent similarities to the concept of organizational ambidexterity (Duncan 1976: 167, Gibson & Birkinshaw 2004: 209). Organizations are surrounded by a variety of tensions (March 1991) and they need to deal with them effectively. March (1991) divides innovation into the categories of exploitation and exploration. Exploration is concerned with leveraging the potential of experimentation, seeking new ideas and generating new items (Andriopoulos & Lewis 2009). Exploitation focuses on the value-maximizing use of resources and abilities (Wadhwa & Kotha 2006). Although these categories appear to oppose each other (Lubatkin et al. 2006) and organizations currently focus on only one of them (Andriopoulos & Lewis 2009), organizations tend to move towards utilizing both aspects to become more homeodynamic. They will approach the edge of order or the edge of chaos.

In order to gravitate around the equilibrium, a homeodynamic organization is constantly fluctuating and oscillating, and, therefore, such an organization is not subjugated to path dependence and the associated lock-in (Sydow et al. 2009). Weick (1976) postulates that organizations are loosely coupled for them as well as their units (or actors, following the terminology of this thesis) to interact freely. Those actors can link with each other at any time and separate if the coupling is no longer needed. Weick (1982) focuses on flexibility, which gives organizations the ability to self-repair. He does state, however, that goals and the dissemination of information are crucial for retaining the loose coupling. Big data can be a contributing force in the diffusion of information and goals within organizations. Nevertheless, such loose coupling will keep organizations flexible and enable the actors to self-organize. In summary, a loosely coupled organization will be able to both fluctuate and oscillate and, therefore, be more homeodynamic.

Although chaos is not intrinsically bad, it can have a major influence on any organization. Too much chaos in an organization may stop it from working. Employees may stop showing up for work due to the lack of a shift schedule, resources may no longer be acquired, products no longer produced, and so on. Any organization requires at least a simple form of order. 1/f noise describes the noise that decreases with an increase of intensity of a phenomenon; there exists a correlation between the noise and the frequency of a certain phenomenon. This 1/f noise or pink noise can be found in many other areas like physical, biological, and economic systems (Bak et al. 1987). Bak (1996:12) asked: "Why are they universal, that is, why do they pop up everywhere?" Here, the focus lies on the observation that "nonequilibrium brings order out of chaos" (Prigogine & Stengers 1984), the discourse of Gaussian and Pareto distribution, and the fact that the effect on organizations can be found somewhere else (Scholz 2013a, Scholz 2015b). Prigogine coined the term 'dissipative structure' and claims that even when an organization faces chaos, a certain form of order will emerge. Dissipative structures will reduce the chaos, thus enabling an organization to be able to sustain itself. Big data contribute to the chaos but can be used to bring order out of chaos. In order to be homeodynamic, however, organizations need to gravitate around the edge of chaos rather than moving towards order.

Yates challenged the influence of DNA on a living system and the idea that "genes act as dynamic constraints shaping product formation" (Yates 1994: 70). For an organization, the analogy with DNA could be the corporate identity (Meijs 2002), the corporate social responsibility (Visser 2011), or the corporate governance (Arjoon 2005). Nevertheless, they are often formulized, institutionalized, and subjugated to several regulations. Consequently, they are more static than dynamic. Following Yates' critique, however, the DNA of an organization may be something different and, furthermore, something more abstract. Especially as DNA follows a certain structure (Watson & Crick 1953). A more dynamic and flexible approach can be found in the string theory in which our universe can be described by only 20 numbers (Greene 2005), "and the wonderful thing is, if those numbers had any other values than the known ones, the universe, as we know it, wouldn't exist" (Greene 2005: 13:27 min). Within those constraints, everything around us has evolved out of these numbers. In a study conducted by Wang et al. (2014), students at Dartmouth were monitored through their smartphones, and the authors identified a *Dartmouth signature*. When ignoring the ethical discussion behind such an analysis, big data become obvious as means of identifying this signature and that, in fact, Wang identified an *organizational signature* (Stein et al. 2016). Organizations are not forced to follow a certain program-driven idea elaborated in their corporate identity or elsewhere, but evolve out of an organizational signature. Nokia, for example, is about *connecting people*. This signature becomes manifest in all their products, from paper to rubber shoes to cell phones. The company stays true to its claim of connecting people. Homeodynamic organizations are influenced by the organizational signature; identifying this signature will be a task for big data, but sustaining it will be a strategic one.

Organizations under homeodynamic conditions need to be able to change and find stability through change (Farjoun 2010). For that reason, enabling them to generate new activity patterns at any time is essential. Activity patterns have the potential of coordinating the functioning of organizations (Lloyd et al. 2001). Lloyd et al. explain that organizations are made up of top-down and bottom-up mechanisms, but "need only small perturbations of their parameters in order to select stable periodic outputs" (2001: 140). Although organizations are able to self-organize at the actor-level and observe both emergent properties and patterns, they will deal with tensions both from within as well as from the outside environment. The concept of self-organized criticality describes the ability of an organization to respond to such tensions. In a homeodynamic organization, the response is quick and economical, which brings up the following difficulty:

- "Too many changes are required at the same time;
- fixing one tension makes another one worse;
- fixing tensions costs money and the firm has no extra funds to spend;
- can't effectively respond to any of them" (McKelvey 2016: 59).

While it is easy to generate new patterns within a homeodynamic organization, the challenge is to keep the organization in balance and sustain its survivability. To summarize this chapter, it can be stated that any of the aspects of homeodynamics are translated into concepts within a homeodynamic organization, as shown in Table 13. However, the organization needs to achieve a homeodynamic balance to stay competitive. Big data will increase the imbalance if left unchecked, but, if used in the right way, big data can make a contribution. The organization now holds new and powerful resources to achieve a homeodynamic balance. As a consequence, big data within organizations will be closely interlinked with the other actors, leading one step closer to a homeodynamic organization.

Table 13: Characteristics of a Homeodynamic Organization

Homeodynamics	**Homeodynamic Organization**
Rate-oriented homeodynamic stability	Dynamic capabilities
Not very far from equilibrium	Ambidexterity
Fluctuating and oscillating	Loosely coupling
Close to 1/f noise informationally	Dissipative structures
Not fixed program-driven systems	Organizational signature
Easy generation of new activity patterns	Self-organized criticality

Translating homeodynamics into a homeodynamic organization is influenced by big data and the core assumptions derived from the impact of big data. Consequently, big data require a more fitting characterization of homeodynamics as well as a shift from a general description of the concept of homeodynamics towards a contextualized description of it within an economic organization. Homeodynamics, combined with the dimensionalities derived from big data observed from an organizational lens, lead to homeodynamic organization as described in this chapter. While achieving a highly homeodynamic organization may theoretically be possible, it will face diminishing returns and the complexity barrier. Trying to be highly homeodynamic adds tension within organizations and will devour resources exponentially. More actors will be involved in keeping an organization homeodynamic and keeping it balanced. That may sound contradictory, as homeodynamics are capable of gravitating and oscillating even far from the equilibrium and far from the chaos. As a result, a homeodynamic organization will be influenced by the constraints already faced by the organization, but big data allow organizations to infuse themselves with new variety and new tools to become more homeodynamic. Big data will not do this on their own, as they are subjective and also tend to standardize. The contribution of big data to the homeodynamic balance depends on the interrelationships between big data (resources) and human (resources) within the organization.

4.2.2 New Roles of the Human Resource Department

Big data will lead to a homeodynamic organization as described earlier and the change can be explained through the implications of the core assumptions of big data. Such a shift in organizational understanding will require a certain reaction from within the organization. Some function will change its role accordingly to tackle the new homeodynamic organizational environment that was triggered by big data and the relation between big data and the people within the organization. One department that already deals with the management of people and, consequently, is involved in change management is the HR department. If big data are seen as a social phenomenon, then the HR department is even more predestined to deal with big data within the organization. However, due to the substantial changes in moving towards homeodynamic organization, the HR department will, at first, react to these changes with the development of new roles for the HR department relating to big data.

The HR function is often the subject of discussion, and many researchers (e.g. Cappelli 2015, Charan et al. 2015, Stone et al. 2015) discuss the role of HRM in the future of organizations. In recent years, the research by Ulrich et al. (2013) is often suggested as a clear picture of how HRM needs to change and what competencies are necessary for HRM to be able to deal with the ever-changing new environment. As Ulrich et al. (2013: 457) state: "HR professionals have often been plagued by self-doubt, repeatedly re-exploring HR's role, value, and competencies", when facing the massive transformation. Technology is seen as a catalyst for the change of HRM function (e.g. Parry 2014), and big data are changing organizations fundamentally, however, somebody will stand up to fill the evolving gap. Big data can be seen as a purely technological phenomenon, but big data will have a stronger impact at the social level and the people within organizations. Consequently, HRM will have the chance to heed the call of big data at these times, and focus on people.

Table 14: New Roles for HR Department

New Roles for the HR Department (Ulrich et al., 2013)	Big Data Specific Roles for the HR Department	Cross-Sectional Role for the HR Department
Strategic Positioner	HR Konstruktor	Big Data Watchdog
Credible Activist	Canon Keeper	
Capability Builder	Theorycrafter	
Change Champion	Built-In Schumpeter	
Human Resource Innovator and Integrator	Data Maker	
Technology Proponent	Data Geeks	

In Table 14, the six new roles suggested by Ulrich et al. (2013) are shown, as are the roles for a HR department concerning big data within an organization. They are following the logic of the roles required, however, with a distinct focus on the special situation of HRM regarding big data. Following the role theory (Mead 1934) it becomes evident that these roles require unique and differentiated positions (Levy 1952), relations within organizations (Parsons 1951), characteristics (Biddle 2013), certain behavior (Linton 1936), a subset of social norms (Bates & Cloyd 1956), and "activities which in combination produce the organizational output" (Katz & Kahn 1966: 179). All of the big data specific roles incorporate this role logic and are derived in order to tackle a certain gap within the homeodynamic organization which aises with big data.

The role of the big data watchdog, however, will be cross-sectional, and as a unification of the HR department with all roles as well as throughout organizations. The big data watchdog is on a higher order for the HR department and, therefore, acts as a guiding system. Thereby it influences the six roles of the HR department for dealing with big data within organizations. Such a role provides the basis for any further changes and modifications caused by the transformation towards a homeodynamic organization.

4.2.2.1 Big Data Specific Roles

In the following, I will briefly describe the six roles on the basis of Ulrich et al. (2013), and afterwards explain the specific characteristics in the context of big data.

HR konstruktor. Strategic positioners focus on the understanding and knowledge of doing business, they need to learn the language of business, contribute to organizational strategy, understand the needs of all stakeholders, and have an intensive knowledge of the business environment of their organizations. Big data will change the HR department in a similar way. The HR department currently focuses on the human role within the organization, however, due to elements like digitization, automation, gamification, and above all big data, the job of the HR department is becoming more technological. Many operational tasks are nowadays performed by software solutions and this will increase in the future. HRM is at a crossroads, as it has space and time to do strategic work and act as designer, creator, networker, and watchdog of the working world within an organization. It can contribute to the strategy of an organization and understand the relational network in which organizations are embedded. Lem (2013) describes such a multifaceted role as "Konstruktor." The role of the *HR konstruktor* is shifting within organizations, to become an integral function that not only looks after the employees but also connects human and machine. This may be a stretch for current HR departments, but it may be part of their survival strategy (Cappelli 2015). If operational tasks are automated and strategic decisions, due to technological complexity, are made by IT, or by quants (Davenport 2013), or data scientists (Davenport 2014), the question arises: What is the necessity of HR? The answer is still the same – to deal with people-related issues – but the embodiment is changing. The HR department needs to learn and understand big data in order to contribute to organizational strategy and have a close look at the stakeholders.

Canon keeper. The next role, of the *credible activist*, is about the credibility of HR professionals and how they build personal relationships and trust. They have a clear message, are trying to improve their integrity, and are experts in business activities. They are also self-aware about their role within organizations. Using big data extensively supports a building of trust within organizations and maintaining this trust over time (Rousseau et al. 1998). The HR department will, therefore, act as a *canon keeper*. Contrary to big data curation, the goal is to become, be, and remain credible about big data use and generate trust concerning big data and the use of big data within organizations. How is big data utilized and in which way? This role is predominantly about showing the actors within organizations that big data are used in a meaningful and positive way by means of communication. Trust can be generated by upholding the integrity and the consistency of big data within organizations. Such a process has similarities to the upholding of a literary canon, as described: "The official canon, however, is sometimes spoken of as pretty stable, if not 'totally coherent'" (Fowler 1979: 98). Part of this canon are "the events presented in the media source that provide the universe, setting, and characters" (Hellekson & Busse 2006: 10). In recent times, canons are part of culture, and emerged as a popular term in, for example, the acquisition of Star Wars: Disney evaluated all stories about Star Wars and categorized them into canonical and non-canonical. Those responsible for this task are called 'continuity cops' or 'keepers' (Baker 2008). Questions about orderliness, story integrity, continuity, internal consistency, and overall coherence are the tasks of such keepers: there are massive amounts of information that need to be integrated, ordered, and made consistent. Big data need to fit with the canons of organizations. Consequently, the HR department will deal with the canonical fit of big data and, in this way, achieve a trustworthy utilization of big data within organizations.

Theorycrafter. The third role of Ulrich et al. (2013) is the *capability builder* in which individual abilities are transformed into organizational abilities. These abilities relate to the strengths of the individual and, consequently, the strengths of organizations, and, therefore, they will influence the organizational culture and identity, or in the terminology of this thesis, the organizational signature. They are, however, concealed within big data when they comprise big data themselves. The HR department, thus, will discover those hidden capabilities at the individual and organizational level. If they are hidden, the HR department crunches the numbers and analyzes the data to discover those capabilities. They act as *theorycrafters*. This term is derived from video games, and describes the search for the optimal strategy within a game on the basis of mathematical and statistical analysis (Paul 2011). Theorycrafters establish simulations that try to mimic the video game and test different constellations on the basis of thousands of iterations. However, it goes beyond the idea of crunching numbers, and is the synthesis of big data analysis and practical experience which can derive usable results. In a podcast from the Training Dummies (2016), the creators discussed the topic of theorycraft in detail and emphasized that it is a combination of theorizing and experience in order to adjust simulation to reality. Some elements are difficult to simulate; others are pretty accurate. However, applying certain metrics to all situations will lead to distorted results, and, therefore, the theorycrafters understand the situation they want to simulate. In the context of abilities, theorycrafters are able to crunch the numbers and apply them to the contextual situation within organizations. Simply mining the data will not be sufficient to identify hidden capabilities; the HR department needs to understand the data, differentiating between signal or noise, and making sense out of the simulations. Finally, the theorycrafter translates the results into action for organizations, and thus can influence the pool of organizational capabilities.

Built-in Schumpeter. The next role is the *change champion*. The HR department supports internal abilities to change, by helping to identify emergent transformations and helping to overcome resistance, and sustain the change ability within the organization. Similar to the tendency to seek a stabilized environment, big data have the tendency to converge, become homogeneous, and favor the mean, though this would lead to statistical errors having a stronger impact (Spiegelhalter 2014), being reinforced over time and becoming difficult to change. The HR department, thus, needs to include a role for stirring up big data within organizations. This role can be called the built-in Schumpeter (Scholz & Reichstein 2015) and people fulfilling this role are trying to continuously conduct creative destruction (Schumpeter 1942). Status quo means deadlock (Farjoun 2010) and is not preferable, however, an HR professional will not destroy in order to destroy, but will have the goal of improving the organization. Big data will be a tool to help the built-in Schumpeter and make organizations more capable of change. The goal is to create alternatives and variety within organizations and within big data. The built-in Schumpeters will evaluate and improve big data use and reconfigure the related investments within organizations. That may be within their own HR department, the HR daemon, or the HR centaur.

Data maker. Another role is the *human resource innovator and integrator.* HR professionals require an in-depth knowledge of HR and acquire knowledge about new trends and new solutions. They are able to translate this knowledge into solutions within organizations, however, Ulrich et al. (2013) emphasize that the HR department focuses on the long-term effects and not on achieving short-term success. In the context of big data, the HR department acts as a type of *data maker.* The term is in analogy with the maker movement (Dougherty 2012) and describes the potential of people to create everything on their own (Stam et al. 2014) and, thus, to create new ideas and new innovations (Lindtner 2014). Big data within organizations depend on such an approach and the ability to think outside the box. Big data will not do such thinking and the HR department will seek out those new ideas with the help of employees, as in hackerspace (Guthrie 2014) or hackathons (Briscoe & Mulligan 2014), and will design a way to integrate such novel uses of big data into organizations.

Data geeks. Finally, there is the *technology proponent.* Technology has increased drastically in recent years and the work of HR is also subject to an increase in technology. Many operative tasks are automatized, other functions are digitized and big data emerges as having an increasing influence on the HR department. Although there is currently an HR-IT barrier, the HR department is driven to overcome it and be open to a more technology-focused HR function. The HR department, therefore, deals with big data, or somebody else will annex this task. This requires some form of cultural change, however, from refusing big data to becoming *data geeks* (Priestly 2015). Although data geeks follow a skeptical approach, they have an interest in utilizing big data in a way that is helpful to employees and organizations. They seek new ways and innovative ideas to analyze the available data and are always looking for new sources of data. Still, their work is within the constraints of the big data watchdog although the HR department is proactively opening up to big data and eradicating the current HR-IT or HR-Big-Data barrier.

In summary, the HR department is facing a big challenge, but it needs to take charge of big data. Big data are not another tool that is delivered or supplied by the IT department or an external business partner. Big data are a critical resource for organizations and will have a strong impact on the work of the employees. Applying big data in the way described will enable the HR department to discover the hidden potential of their employees and generate a competitive advantage for organizations. Big data allow the HR department room to focus on the strategic perspective of improving and helping employees to improve. Although there are self-doubts and people are constantly re-exploring the HR role (Ulrich et al. 2013), big data offer an opportunity to assume a strategic and integral role within organizations and influence their survivability.

Milan Lab

In order to understand the new role of HRM, it is useful to look into sports again. Davenport describes the interest of HR professionals in sports as the following: "Still, sports managers – like business leaders – are rarely fact-or-feeling purists" (2006: 102). I have talked in this thesis about Oakland Athletics and FC Midtjylland, but there are many other sport teams, for example the football clubs TSG 1899 Hoffenheim in Germany and Bolton Wanderers in England, that highlight the extensive use of data. One example in particular that seems strikingly fitting is the Milan Lab of the Italian football club AC Milan. The club is using modern technology extensively to improve health quality and "predict the possibility of injuries" (Kuper 2008). Interestingly, Meersseman, a former director, compares the lab to a car dashboard and players to drivers: "There are excellent drivers, [...] but if you have your dashboard, it just makes it easier" (Meersseman in Kuper, 2008). Big data support the work of a coach and their staff.

There seems to be a focus on body health issues and a focus on data-driven decisions, however, the Milan Lab tries to improve the soul of the players as well. For example, if they have had traumatic experiences, like the brutal injury of Schewtschenko, the Milan Lab and the staff help the players be able to deal with their fear (Biermann 2007). Although it is difficult to quantify the effect of the Milan Lab, it seems that there is a positive effect. Players are able to compete at an international level at older ages (Newman 2015a). The team won the Champion's League in 2007 with an average team age of 30 years, and Paolo Mandini, the captain of the team, was 38 years old (Transfermarkt n.d.). This is interesting in times of a general 'youthism' in football (Grossmann et al. 2015). Big data change the role of the coaches and the staff.

Big data are a source with which to improve work, however, they do not make work magically better, and people are still essential. This can be seen in sports, and will be seen in organizations as well. Currently, many are praising the potential of predictive policing (Beck & McCue 2009), but the advantage of predictive policing is not that crimes are discovered by algorithms, but that the police can do their work faster, more systematically, and more efficiently (Peteranderl 2016). Again the role of the police officer has changed.

This will be similar to the use of big data within organizations. Big data will enable people to become more efficient, but it is the HR department that makes big data and the people more capable of dealing with each other. The role of the HR department will drastically change, however; it will be responsible for exploring new potentials and new ideas for the use of big data. Big data do not magically make organizations better places, and people are still greatly involved. This unique use of big data will be a competitive advantage, and such a unique use comes from the people involved. The Milan lab will not share their information (Newman 2015a), and FC Midtjylland has no interest in sharing their secrets either (Biermann 2015). Competitive advantage is created by people and not by big data, consequently any organization requires a unique way of using big data and not buying 'off the shelf' tools from some external provider.

4.2.2.2 Big Data Watchdog as Cross-Sectional Role

Those different roles are dealing with many facets of big data within organizations, however, they can be seen as relatively separated from each other. It is essential to have some sort of cross-sectional role for the HR department since using big data within organizations leads to complex interplay and interaction; big data needs to be supervised within organizations. Due to the contextualization of an organization in particular, and, subsequently, the organizational signature, only a portion of big data is useable for organizations. Big data will also have an influential impact on organizations, therefore, big data are closely supervised and organizations are capable of dealing with big data. Therefore, the HR department will be authorized to watch over big data.

Such a role is comparable to that of a watchdog. The term "watchdog" is currently being discussed in pop culture, largely due to the video game "Watch Dogs" (issued by Ubisoft in 2014). The game is located in a futuristic Chicago which is under total surveillance that is misused by the developer (a company) and the users (the city and the police). The protagonist also has full access to personal data of all inhabitants and acts as a watchdog and vigilante. This mirrors the recent discussion of big data and the role of the NSA (Gallagher 2013) and, therefore, reflects the current zeitgeist, however, such a watchdog is not only essential at the societal level, but especially at the organizational level.

Merriam-Webster's dictionary defines a watchdog as "a person or organization that makes sure that companies, governments, etc., are not doing anything illegal or wrong". It is often used in the context of investigative journalism, where the journalist acts as corrective (Miller 2006, Rensberger 2009), as well as other non-profit organizations that act in a similar way (Rao 1998). In recent times, and especially since the case of Edward Snowden, a watchdog has been compared to a whistleblower (McLain & Keenan 1999), however, external whistleblowing is seen as a last resort (Miceli & Near 1994), while a watchdog intervenes at an earlier and still changeable phase, fulfilling an important control function. Being a watchdog is, therefore, not just protecting, but also guiding, and acting in general as a sort of corrective within an organization.

As big data are always connected to humans, it may seem obvious to give the role of watchdog to HRM. In order to provide justification for the proposal that the HR department is an appropriate office to handle this watchdog responsibility, the functional role of HRM will be highlighted. Modern HR departments already go beyond the stereotypical "hiring and firing" and focus on employability. In times of increasing employee participation (Busck et al. 2010) and its visibility in employer branding (Wilden et al. 2010), the HR department is dedicated to the task of acting in the interests of employer and employee at the same time. It is a "first and second party" rather than a "third party" and, therefore, has internalized the ethics of both the company and employees. In fact, if the goal is to help improve the performance of employees and the relationship between employer and employee, a department responsible for workplace training is a good fit for this watchdog role, and the HR department usually covers this responsibility. This description also applies to their unique and differentiated position within organizations.

Specialists in employment law can be found in the HR department, being able to resolve legal questions. Since the use of big data is a multifaceted issue, a consortium of employees from various departments and representatives from trade unions and work councils could be integrated in a steering committee headed by the HR department. Although a suitable legal landscape is missing at the moment, or at least is underspecified in many cases, the HR department can base its decisions on its broad experience with other sensitive and more deeply regulated issues such as diversity (Zanoni & Janssens 2004). In analogy to this, the HR department has already exercised restraint in going beyond the boundaries of "good corporate governance" (Fauver & Fuerst 2008: 673). HR departments are following a certain subset of social norms and behave in a distinct way.

Another element is that the modern and digitally competent HR department is able to provide its professionalism regarding big data handling. This role includes ensuring that the analytics of the big data are correct, unbiased, and not taken out of context (Kitchin 2014a). It also includes ensuring the appropriate retention of data, making sure that the data are used legally and will remain internal and secured, and maintaining transparency in data collection and accessibility for appropriate parties. On the one hand, the HR department is bound to secrecy and, therefore, required to keep internal information internal. On the other hand, the HR department needs to bring together big data experts, increasing their expertise in IT, the law, and analytic skills, thus training them to handle this responsibility. With such a diverse set of characteristics and with a unique position within organizations, the HR department contributes to organizational output.

These elements emphasize the claim that HRM could act as a big data watchdog; however, they also highlight the complex situation of big data use within an organization and the essential need to keep big data within organizations. Such a watchdog needs to deal reactively with the impact of big data on an organization from a social and ethical perspective. This role is highlighted in most of the current criticism concerning big data (e.g. Boyd & Crawford 2012, Kitchin 2014a).

Target Baby versus One Family Baby

In order to explain the uniqueness of the big data watchdog as well as the need for such a role, the following examples will present two different approaches to dealing with information derived from big data. In 2012 a debate erupted due to an incident involving Target (Hill 2012). The company has extensive information about their customers and their buying history, such as baby-shower wish-lists that can be organized through Target meaning that Target knows which customers are pregnant. Using these data sets, they derived a pregnancy prediction score. One sign used to identify pregnancy among their customers is the following:

> "Many shoppers purchase soap and cotton balls, but when someone suddenly starts buying lots of scent-free soap and extra-big bags of cotton balls, in addition to hand sanitizers and washcloths, it signals they could be getting close to their delivery date" (Duhigg 2012).

Target utilized this information and sent the customers they believed to be pregnant coupons for baby clothes. Hill (2012) described a case in which a customer showed such a coupon to the Target manager addressed to his daughter. Although the manager apologized, it turned out that the daughter was pregnant, but hadn't told her father. Remember this is a real case, in which Target discovered a pregnancy before a close family member, and that actually it was through a relatively simple method (Ellenberg 2014). From a marketing perspective, it makes sense to use the information, but from the corporate social responsibility perspective, it may harm the company's credibility if customers perceive it as unethical (Schramm-Klein et al. 2016).

A similar example involves the episode "Connection Lost" of One Family (Season 6 Episode 16). This time, the mother tried to talk to her daughter after a fight couple of days ago and wanted to know what was wrong with her daughter. Eventually, the mother was stuck at an airport and only had access to the internet, so she phoned relatives by using a video messaging platform, viewed the Facebook feed of her daughter, and accessed her daughter's iCloud by hacking her password (it was an easy password). A package from Amazon arrived at the home and was opened by the father. She gathered the following information. Her daughter recently married (Facebook), had posted several pictures with a male friend (Facebook), is at the moment in Las Vegas (Find my IPhone), and had ordered baby books (Amazon). This all suggested that the daughter was pregnant and had eloped, however, at the very end, the daughter had just fallen asleep and there was nothing to worry about. The daughter changed her Facebook status as a joke and had met with a friend who borrowed her phone and the books were for her boss. The daughter screamed afterwards: "Borders, Mom!"

Although this example is about families, organizations have likewise the ability to access a similar amount of social data. In times of the dissolution of labor, especially, there is no difficulty for organizations to access such data. Are there changes in behavior, does someone stay longer at work, how much coffee do they drink, how much do they talk with others? Is it ethical for an employer to check the social media profiles of employees if they call in sick? In a survey by Jobvite (Singer 2015), they discovered that only four percent of recruiters do *not* use social media. Consequently, recruiters will screen social media profiles and potential employees will clean up their profiles (Brown & Vaughn 2011). But the overarching question remains: Is it ethical to use such information?

Using big data will always have an ethical dimension and the answer to the ethical question will vary from organization to organization and with their surrounding network and stakeholders. It is essential to deal with this question carefully. Big data are volatile and new ways of generating big data are emerging; consequently organizations need to be watchful and vigilant about new developments. The aim for any organization is to find a fitting answer for their organization. Some will find what Target did acceptable, and many would be outraged by the action of the mother, but both examples used big data to gather information.

4.2.3 Human Resource Daemon

The previous chapter described the reaction of the HR department to the transformation towards homeodynamic organization triggered by big data. However, this is only the first step in dealing with big data within an organization. Only reacting towards big data will not be sufficient, consequently, the HR department will establish new structures in order to use big data in a valuable way for the organization. The HR department now has the competence to deal with big data and, furthermore, there is the need to create an environment for big data within organizations, as big data potentially have a life of their own which would cause a loss of control in the organization. Big data, however, influence the organizational network and all actors within organizations. Such an independent existence of big data would put them into a black box and organizations would treat big data as an external influence. In order to utilize big data to the fullest and use them for homeodynamic balance, organizations need to understand organizational big data and integrate big data into the organizational network.

Letting big data roam freely through organizations is not an alternative path, as big data will overburden individual actors within organizations. It is, therefore, necessary for an institution to deal with big data and be accountable for big data within organizations. Big data are driven by technology and, currently, big data are often deliberately used under a veil of ignorance. Although there may be reasons for putting big data into a black box, they need to be contextualized within an organization. This understanding can only be achieved by adding the social perspective to big data and not solely focusing on technology. Big data are also not objective and so in order to use them in an organizational context, they require transformation. Somebody will also utilize, distribute, and train the employees. Big data within organizations are, therefore, every employee's business. I propose that the human resource department is the most suitable candidate for dealing with big data within organizations. Although HR departments probably lack big data knowledge and the competencies to handle them, they are capable of dealing with people. This seems to be exactly what is needed:

> "The challenge is not just a technological one: the selection, control, validation, ownership and use of data in society is closely related to social, ethical, philosophical and economic values of society" (Child et al. 2014: 818).

It sounds somewhat paradoxical, but big data emphasize the "people question" within organizations. Big data require people within an organization and will not replace them. Using the Moneyball example, Silver (2012) describes the time after the Moneyball incident as a fusion between worlds: that statistics and scouting might work together and mutually achieve more than using either data or people. He claims that this synthesis helped the Boston Red Sox to win their first championship title in 86 years. There are various other examples, one of which being chess (Kelly 2014, Ford 2015), that reveal the advantages of this mixed usage. It is therefore no longer true that data will create an competitive advantage as everybody has access to them; in a world driven by data, people will make the difference. Those people need help from an HR department that moves beyond operational tasks and works towards becoming a strategic partner (Charan et al. 2015) and the *social* big data expert capable of handling big data.

In order to deal with big data, it is essential to integrate them into organizations. This implementation does not lead to a complete transparency of big data which would overburden employees, but enables the HR department to utilize big data to their fullest capacity. Most current software, and many programs and applications, are separated into a backend and a frontend. The backend is the system that operates in the background and is invisible to the user. The frontend is the user interface (UI), which is visible. A user in this case is an employee in the organization. The HR department does the heavy lifting of big data in the backend, and designs a frontend for the employees. A backend is sometimes called a 'daemon' which is why I propose a system called the human resource daemon in analogy to the Laplace daemon. Although the HR daemon will not provide all answers, the goal is to outline a hypothetical organizational daemon that is capable of giving solutions to all questions within organizations.

Such a daemon and, subsequently, the HR department, will deal with three aspects of big data. Firstly, big data will be generated. I use the term 'generate', as big data are always modified in one way or another when entering organizations. At least, any external big data are labeled 'external' and any internal big data are labeled 'internal'. This differentiation will be judged in a certain way and will influence organizations in different ways. Like any analysis, the HR daemon constantly generates new big data all the time. Secondly, big data will be evaluated. This evaluation already starts in understanding the source of the information. Information, for example, has a half-life and using old information comes at a risk. Another concept that deals with evaluation is the categorization of high data swiftness and high data rigor. Finally, there is the aspect of monitoring. Big data will influence social interaction within organizations and have an impact on the organization. The HR department establishes structures to watch over the influence of big data on the organization. Generating big data will be conceptualized in the *data farm*, evaluating big data in the *fog of big data*, and integrating big data in the *big data immersion*.

4.2.3.1 Data Farm

Big data construct a certain form of reality, be it influenced by a social constructivism or on its own through a data constructivism. Seen from a temporal perspective, however, this reality will be reinforced over time. This type of reality is fortified, and big data act self-referentially towards the acceptance of such a reality. Big data are trapped within a self-reinforcing circle which leads to a risk of uniformity, an increase in homogeneity, similarity, and convergence. This could be beneficial if big data were to be objective, not in the light of their subjectivity. Consequently, big data distort reality towards one potential and subjective reality. This tendency to reduce variety and, therefore, reduce big data, is explained in the following example concerning maps:

> "A good map eliminates as much spurious information as possible, so that what remains is just enough to guide our way. Moreover, when the map is well made we gain a deeper understanding of the world around us. We begin to recognize that rivers flow in certain directions, towns are not randomly placed, economic and political systems are tied to geography, and so on" (Miller, J. H. 2015: 1).

Although it makes sense to eliminate spurious information, the problem with big data is that spurious and relevant information is indistinguishable. Navigating by means of a map is a simplistic goal and unnecessary information is easily singled out (e.g. Miller, J. H. 2015), but achieving homeodynamic balance within a turbulent field is an obviously abstract goal and relevant information can become dynamically irrelevant and vice versa. It is, thus, essential to gather as much information as possible. Assuming big data have a tendency to destroy variety, the HR department creates it.

The concept of the data farm aims at creating variety in order to generate more big data and, most importantly *more diverse* big data. In today's age of technological advancement, storing huge amounts of data has become very affordable (Murthy & Bowman 2014). In the context of an organization, only a small portion of that big data is important. The first way of generating big data involves using a variety of algorithms. If several algorithms are available, all are applied. There is no objective explanation of why one algorithm may be superior to another; some have an inherent ideology (Mager 2012), and others seem to be correct even though the programmer does not understand why (LaFrance 2015). To ensure the potential of making a decision, a selection of big data is required.

Big data are also learning through machines, and people learn as well, so the data farm will learn and will use new and different algorithms. Although there are often good reasons for choosing a certain path at a bifurcation, under homeodynamic conditions, it may be a total evolutionary dead end. The next task for the data farm is, thus, to archive evolution (Scholz 2016b). It is possible that an earlier evolution of the data farm may be more accurate regarding newer changes in the environment. Above all, it has become more evident in recent research in organizational theory that history matters (Sydow et al. 2009). Remembering history helps understand recurring patterns (Turchin 2008) and can be used to increase the ability to predict events (Spinney 2012). Although a precise prediction of the future may be impossible, any organization has the tendency to tackle situations in a certain way, depending on their organizational signature. Such a data farm is shown in Figure 9. Every evolution and every algorithm creates a new data stream and a self-consistent form of reality. This can be compared to the idea of the multiverse (Deutsch 2002) according to which there is an infinite number of universes, each of which differs slightly from all others. The authors of the science-fiction work "Long Earth" (Pratchett & Baxter 2012), propose the possibility of there being an infinite number of different Earths. Human characters in the book are capable of switching between these Earths. Some multiverses are only marginally different, others (the authors call them jokers) vary drastically. This can be seen as a parallel to the data farm and accentuate the fact that deviations of any kind will have an influence.

Figure 9: Evolution of Data Streams within the Data Farm

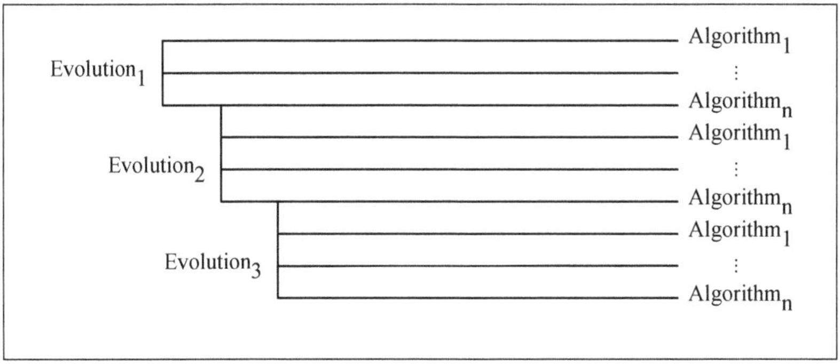

The general goal of a data farm is to increase the variety of big data within organizations, thereby counteracting the tendency of big data to destroy variety. Such an increase in big data can add to their overall preciseness. The HR department can paint a more granular picture of available information. This is interesting, as there is a concept in cryptography and collective intelligence which states that "no information is information" (Grimson 1980: 114). The absence of certain information will tell a story, and knowing that all information is utilized will increase the story.

Finally, the data farm adds a certain scalability to big data analysis within organizations. It is a tool that everybody can use. Every new analysis, however, is added as a new data stream to the data farm and acts as a data mutation or a 'joker'. Those new data streams are highly contextualized and include all the relevant metadata required to understand what properties have changed. This form of scalability is essential for the fog of big data. Nevertheless, it is important to highlight that the data farm remembers any form of data mutation, and can use this for all big data analyses if needed or required.

Psychohistory

When talking about big data and predictive analytics, people (e.g. Turchin 2012) often cite Asimov's "Foundation" (1951) and take it as an example that everything is predictable. That is, however, too overenthusiastic, and in the course of the book Asimov reveals that predicting the future is a very complicated task. At the demise of a Galactic Empire, Hari Seldon, the creator of psychohistory, created two Foundations and established the Seldon Plan.

> "The Seldon plan is neither complete nor correct. Instead, it is merely the best that could be done at the time. Over a dozen generations of men have pored over these equations, worked at them, taken them apart to the last decimal place, and put them together again" (Asimov 2010: 497).

It seems impossible to predict the future. By predicting the future in this particular way, however, Seldon influences the future, as the one known Foundation is an enclave of knowledge for physical sciences. Similar to the uncertainty principle (Heisenberg 1927), people will attempt to make this *particular* future a reality which is the task of the second and hidden foundation consisting of social scientists. Their task is to manipulate society to follow the Seldon plan.

In order to keep society on track, this second foundation adds new amendments to the plan and influences society in a certain way. They do, however, calculate for all eventualities and especially for all unknown unknowns (Pawson et al. 2011) as shown in the character called Mule in the book. The second foundation derives a variety of amendments and plans, and administers numerous changes to bring society back on track. The assumption is that it is the social scientists who utilize the Seldon Plan, proactively trying to make it happen, and that the physical scientists subordinate themselves under the Seldon plan. Similar behavior can be observed in the use of big data (e.g. Lange 2002).

This example illustrates the importance of both keeping an eye on everything, and increasing the variety of data streams, enabling any organization to apply changes to the system. Psychohistory is linked to the field of cliodynamics (Finley 2013). This field tries to analyze history quantitatively in order to discover patterns that can be used to predict the future. Although Turchin (2008) limits the possibility of using history or any other way of predicting the future due to "mathematical chaos, free will and self-defeating prophecy" (2008: 35), cliodynamics can be used to learn lessons and discover empirical regularities (Turchin 2012) or remember the mistakes of the past (Scholz 2016b). Psychohistory and cliodynamics depend on big data, however, and depend on somebody separate to curate the different data streams.

4.2.3.2 Fog of Big Data

One issue that is crucial for the use of big data is their possible incorrectness. Information from big data can be outdated, collected for a different purpose, be tampered with, be incomplete or fragmented, or be faulty due to measurement errors or errors in communication (be they technical or due to human communication). Using big data comes at a risk due to the uncertainty of the value of information generated. In military terms, and within current video games, such strategies for dealing with uncertainty are called the "fog of war". The term was first used by von Clausewitz, using the terminology of *Nebel des Krieges* and he describes it as follows:

> "War is the realm of uncertainty; three quarters of the factors on which action in war is based are wrapped in a fog of greater or lesser uncertainty. A sensitive and discriminating judgment is called for; a skilled intelligence to scent out the truth" (von Clausewitz 1832/1976: 101).

The concept of fog of war is analogous to the use of big data which is why I propose establishing the concept of *fog of big data*. Big data are not seen as a reliable source without sufficient information about them. Metadata will, therefore, become increasingly important for the HR department, as well as in the social media appearance of employees. Obviously the actions of employees on these platforms say something about them. It is common knowledge, however, that companies monitor people on social media, and so potential employees clean their social media profile. Regardless of legal questions raised by this social media research (Hoeren 2014), recent developments in research question the reliability of data collected this way (Brown & Vaughn 2011). People adapt to the use of big data within organizations in a certain way and so organizations will deal with the fog of big data and "scent out the truth", as von Clausewitz (1832/1976: 101) explains.

Dealing with the fog is a strategical task and requires an active use of resources. As in military strategy, the HR department needs to actively scout for reliable information and evaluate existing information. There is also a need to evaluate the tradeoff of having many risky data points and only a few precise ones. The HR daemon, therefore, requires a range of tools to deal with big data. One tool is to give organizations the ability to identify faulty data by means of *big data baloney detection*. Another tool serves the purpose of simulating all potential outcomes of thinkable and *un*thinkable strategies, as described in the concept of *big data tinkering*. Using the first tool densifies the fog of big data as it depends on rigorous analysis. Only a smart portion of big data will be visible, but the picture will be clear. The second tool will cause only a light fog of big data as it depends on outside-the-box thinking. The picture will be fuzzy, however.

The fog of big data reveals the potential of big data within organizations, and also the uncertainty and risks concerning big data. They will be evaluated from all perspectives which is why the comparison to the fog of war seems fitting. Somebody derives a plan and all available information needs to be evaluated and calculated. Power and knowledge are not purely derived from information, but from its translation into action. The fog of big data strengthens the previous claim that the competitive advantage will be found within the human actors. Algorithms are bound by their rationality and their rules, then the human factor (Zuboff 2014) adds irrationality and diversity into the mix. Both big data and people, therefore, represent sources of risk, but big data will be shackled within people's subjective reality, and will end up as a fog of big data with which people will retain the ability to interact dynamically.

4.2.3.2.1 Big Data Baloney Detection

We are surrounded by data and currently big data are being put into a black box and perceived as something magical. This observation, as stated earlier, puts organizations into a difficult position. Sentences like "the data clearly states" or "there is a significant correlation" are common and emit confidence, maybe even faith in big data (Boyd & Crawford 2012). Nevertheless, big data do not increase the precision of data analysis; on the contrary, big data increase the veil of ignorance (Rawls 1971) and people's trust in numbers (Porter 1996). Similar to the claims that big data eradicate theory (Anderson 2008) and that, they are, in fact, strongly theory-driven (Mayer-Schönberger 2014), big data do not lead to more precise observations, but much rather increase the number of observations that are plausible at first sight but often turn out to be wrong.

The topic of dealing with observations and the potential of incorrect observation is discussed in great detail by Popper (1959), introducing the principle of falsifiability. He claims that there is a general asymmetry in analyzing hypotheses. Although it is not possible to verify a hypothesis in its totality, it "can be contradicted by singular statements" (Popper 1959: 19). Big data strengthen the claim of falsifiability, but the opposite effect is currently observable. A large group of people is content with discovering patterns or correlations within big data and believes that this is sufficient due to the amount of big data available. It seems that big data are subjugated to economies of scale to which the problems and the errors in big data are also subjected. As Spiegelhalter reasons: "Serious statistical skill is required to avoid being misled" (2014: 265).

Sagan (1996) was facing a similar situation when he discussed ways to deal with pseudoscience. He demonstrated that there are ways to identify solid scientific research and rigorously tested work and not fall for poorly conducted research. He developed a 'baloney detection kit' that equips people with the tools for skeptical thinking, an ability more crucial than ever as big data have become so complex that they may in fact appear as magic. Many of the tools proposed are perfectly suited for the use of big data. Sagan proposes the following nine tools for his baloney detection kit:

- "Wherever possible there must be independent confirmation of the 'facts'.
- Encourage substantive debate on the evidence by knowledgeable proponents of all points of view.
- Arguments from authority carry little weight – 'authorities' have made mistakes in the past. They will do so again in the future. Perhaps a better way to say it is that in science there are no authorities; at most there are experts.
- Spin more than one hypothesis. If there's something to be explained, think of all the different ways in which it *could* be explained. Then think of tests by which you might systematically disprove each of the alternatives. What survives, the hypothesis that resists disproof in this Darwinian selection among 'multiple working hypotheses' has a much better chance of being the right answer than if you had simply run with the first idea that caught your fancy.
- Try not to get overly attached to a hypothesis just because it's yours. It's only a way station in the pursuit of knowledge. Ask yourself why you like the idea. Compare it fairly with the alternatives. See if you can find reasons for rejecting it. If you don't others will.
- Quantify. If whatever it is you're explaining has some measure, some numerical quantity attached to it, you'll be much better able to discriminate among competing hypotheses. What is vague and qualitative is open to many explanations. Of course there are truths to be sought in the many qualitative issues we are obliged to confront, but finding *them* is more challenging.
- If there's a chain of argument, *every* link in the chain must work (including the premise) – not just most of them.
- Occam's Razor. This convenient rule-of-thumb urges us when faced with two hypotheses that explain the data *equally well* to choose the simpler one.
- Always ask whether the hypothesis can be, at least in principle, falsified. Propositions that are untestable and therefore unfalsifiable are not worth much. Consider the grand idea that our Universe and everything in it is just an elementary particle – an electron, say – in a much bigger Cosmos. But if we can never acquire information from outside our Universe, is the idea not impossible to disprove? You must be able to check assertions out. Inveterate skeptics must be given the chance to follow your reasoning, to duplicate your experiments and see if they get the same result" (Sagan 1996: 210–211).

On the basis of this baloney detection kit, I will present a *big data baloney detection kit*. This kit will be a helpful tool for the HR department to discover faulty data as well as faulty conclusions and create a structure in the HR daemon that tackles the veil of ignorance in organizations. The kit consists of: (1) The necessity to find other data sources to seek validation. Big data are subjective but big data represent a way to access other sources without any effort. Although a certain data set may reveal facts, they are always checked and validated. (2) Big data analyses are not performed by only a few people or only certain people or departments within organizations. Everybody who is influenced by the results of a certain big data analysis needs to have a voice. Many big data decisions will involve employees in one way or another, that is why the HR department and the works council are part of them. (3) Big data are subjective and even data from authorities like government agencies will be distorted in some way. This may happen on purpose or by mistake, but without precise knowledge about the way in which data are collected. These data are not superior simply because they were collected by an authority. (4) In the context of big data, the hypotheses surrounding such correlations derived from data mining will become ever more important and have a major impact on the use of big data. Although some correlations may lack all logic (Vigen 2015), there are many correlations that appear, seem to make sense. These correlations may discover a causal effect; nevertheless, correlations do not give information about causal relations. Who influences whom? Correlations in big data can often be explained in some way, but using multiple explanations will at least lower the chance of choosing the wrong one. (5) If big data reveal correlations, the explanation behind each one becomes more relevant and can be a source of criticism. Although a correlation makes sense in the subjective reality of one person, it may be baloney in other subjective realities. (6) In terms of big data, to quantify does not mean to use more data, but to evaluate the quality of the data available. Although some data may be more numerical, they could be of poor quality. Quantity are, thus, replaced with quality. Good data always trump bad data, however, the answer is not that easy if it comes to the comparison between good data and many data. As Hand (2016: 631) states: “Large does not necessarily mean good, useful, valuable or interesting”. (7) Big data are always a mosaic of different data sets (Sprague 2015) and in order to improve the big data analysis, every source and every link is checked for quality and for potential biases within the data set. (8) Occam's razor can be applied in the same way and the emphasis lies on explaining the data set equally well. (9) Results from big data are always tested for errors. Big data analyses need to become more transparent in order to understand their reasoning (Dalton & Thatcher 2014). It is understandable that many big data analyses cannot be duplicated. Big data are closely interlinked with the source as well as with the results (Ansolabehere & Hersh 2012). Many algorithms are self-learning and evolve over time, but in order to discover baloney, people possibly need to follow skeptical reasoning.

Big data have another serious problem. Hypotheses are no longer formulated in advance. Big data can find patterns and correlations without requiring any hypotheses at all. Organizations face the problem of HARK (Kerr 1998), the acronym for the practice of hypothesizing after the results are known. Although it may not have a strong influence on research, as those hypotheses are still rooted in preliminary literature review and research (Bosco et al. 2015), the appeal of HARK in an organization is much greater and may lead to falsifying data (Kerr 1998). This phenomenon helps describe the problems when faced with big data. In most cases, organizations will conduct HARK on a large scale. HARK seems to be considered a legitimized method (Loebbecke et al. 2013). Given that it is not an unsound method per se, dealing with HARK requires different tools. The big data baloney detection kit will contribute to the competences of organizations in minimizing the potential for HARK to have a negative impact.

Sagan defines the following fallacies that are to be avoided as part of the baloney detection kit: "ad hominem, argument from authority, argument from adverse consequences, appeal to ignorance, special pleading, begging the question, observational selection, statistics of small numbers, misunderstanding of the nature of statistics, inconsistency, non sequitur, post hoc ergo propter hoc, meaningless question, false dichotomy, short-term vs. long-term, slippery slope, confusion of correlation and causation, straw man, suppressed evidence, weasel words" (Sagan 1996: 212–216). I will not describe them as many of the fallacies are already described through cognitive biases and statistical errors.

Putting big data into a black box and seeing the work of big data as magical would have prompted Sagan to categorize big data as a source of nonsense. Implementing a tool for skeptical thinking into the HR daemon is essential. The big data baloney detection kit consists of general elements with which to understand the reasoning behind any big data analysis and will help open up the black box. Intensive baloney checks, however, will require a vast amount of resources which render the intensity of use of such a kit a strategical decision, especially considering the tradeoff of potential risks.

Where is everybody?

In his book, Sagan (1996) addresses the possibility of alien life and tries to debunk the potential of alien abduction. Although the discussion about alien abduction is indeed pseudo scientific, the question of alien life is a fitting example with which to describe the relevance of the big data baloney kit. Big data and many new telescopes allow astronomers to gain a more precise picture of our universe than ever before. Researchers have discovered over 2,000 exoplanets since 1988 (http://exoplanet.eu/catalog/). The universe is vast, however, which is why our knowledge of it is massive and minuscule at the same time. With all this information, we cannot answer the question of whether or not we are alone in the universe.

Scientists struggle with the so-called Fermi paradox (Webb 2002). The universe is so vast and old, and there are so many earth-like planets that there is likely to be extraterrestrial civilizations and they may have visited Earth. Webb (2002) suggests fifty answers to the question. He proposes that the following scenarios are the most plausible: that they are signaling but we do not know how to listen and we do not know at which frequency to listen, and that we have not listened long enough. There are many solutions to the paradox. Scientific observations, however, are generally faced with rigorous skeptical thinking.

In recent years, the planet KIC 8462852 has been the source of vivid discussions. This planet was behaving unusually. Boyajian et al. (2015) discussed a variety of scenarios that could describe this planet. In addition to those scenarios, they do not rule out the possibility, although extremely small, that it was built by somebody. "Aliens are always the very last hypothesis you consider, but this looked like something you would expect an alien civilization to build" (Andersen 2015). Wright et al. (2016) agreed with the statement that explaining such an observation with alien life is the last resort, especially as the hypothesis cannot at the moment be disproven. Explaining inexplicable phenomena induces an "aliens of the gaps" (Wright et al. 2014: 3) fallacy. Ultimately, it is more probable that there is just noise in the data rather than signs of alien life (Boyajian et al. 2015, Wright et al. 2016).

Although we are currently drowning in astronomical data (Zhang & Zhao 2015) and astronomers continue to find new planets and new phenomena, they are using both the classical baloney detection kit and the big data baloney detection kit. Researchers want to eliminate all possibilities before claiming the discovery of alien life.

> "With this in mind, it's possible that this binary nature is due to scientists being extra cautionary on how they present results to the public. If something extraordinary such as life beyond Earth is detected, then we'd better be prepared to unequivocally back up such a statement" (Boyajian in Greene 2016).

Such an example highlights the relevance of being precise and cautious with this particular topic. However, organizations also need to be cautious with their use of data. Using data that may or may not be accurate and being satisfied by correlations, or the first hypothesis that comes to mind, will harm organizations. Critical decisions with far-reaching consequences need to be handled in a deliberate and precise way, just as much as the question of whether or not we are alone in the universe.

4.2.3.2.2 Big Data Tinkering

By detecting baloney within big data, organizations are in danger of creating a tunnel vision. Organizations will restrict themselves within the possibilities of big data. This may even slow down organizations or lead to a dead-lock (Takebayashi & Morrell 2001). The strength of big data is that it is possible to just look into the data, to find patterns, discover coincidences nobody even thought of, or simply simulate someone's crazy idea. An organization needs the ability to "play around" (Jacobs 2009: 36) with big data and to have space for exploratory analyses.

Big data alone will not be a source of creativity or innovation, but will enable people to think outside the box and will augment people with new tools. A potentially fitting terminology for this is 'the bricolage' by Lévi-Strauss, who describes it as "doing things with whatever is at hand" (Lévi-Strauss 1966: 17). In an organizational context, bricolage is often linked with entrepreneurship, innovation, and organization theory (Duymedjian & Rüling 2010) and tackles resource allocation within an organization or, to paraphrase, the process of "creating something from nothing" (Baker & Nelson 2005: 329). Although the analogy 'from nothing' is unclear within an organization, as some resources will be re-allocated, those resources, in this case often people, will be used in a different context and environment. Organizations are forced to improvise, fixing things, or designing new things (Weick 1993, Louridas 1999). Weick describes the need for such bricoleurs and their ability to improve and redesign organizations with the following reasons:

1. "People are too detached and do not see their present situation in sufficient detail;
2. past experience is either limited or unsystemized;
3. people are unwilling or unable to work with the resources they have at hand;
4. a preoccupation with decision rationality makes it impossible for people to accept the rationality of making do; and
5. designers strive for perfection and are unable to appreciate the aesthetics of imperfection" (Weick 1993: 353).

These reasons are strengthened by big data baloney detection and shackle people within organizations by focusing on the past, the existing observations and the tendency to implement structures within organizations. They do, however, thereby increase the rigidity of organizations, implement a strong lock-in (Sydow et al. 2009). There may be a tendency to stabilize organizations (Weick 1979), but in today's world stability means being dynamic (Farjoun 2010). Consequently, the work of a bricoleur seems to be an efficient way to increase the dynamization within an organization.

Although bricolage may be a fitting description, the term is sometimes used in the sense of errors, or shoddy piece of work. In my opinion, a more precise term is *tinker*. The Merriam Webster dictionary defines the term as somebody who is repairing and working in an *experimental* way. This is still similar to the bricoleur of Lévi-Strauss, but evades the negative connotation of the word in the English language. In the video game "World of Warcraft", tinkerers are described as:

> "The creators of incredible inventions from steam saws to siege engines, their devices allow them to overcome nearly any situation – and if they don't have the device they need, they just might be able to design and create a new one on the spot" (Kiley 2005: 86).

Tinkerers are known for using their resources at any time and any place. I suggest that big data tinkering will become an essential element in the use of big data within organizations. Creating new ideas and new concepts, using tools in different ways, and utilizing big data for such tinkering or for innovation will boost the competitiveness of any organization. Such tinkering is not blindly mining big data driven by data, but combines the creativity of people with the computational power of big data. Big data are shackled by their rationality, although this is distorted in some ways, and by their boundaries. Tinkerers can add their irrationality to the mix and drastically expand the benefit to be gained from big data. 3D printing and the maker movement (Dougherty 2012) can be used as an example. Although a 3D printer can print everything, somebody tells it what to print. There are almost no restrictions (e.g. food or steel printing), but a tinkerer needs to use the 3D printer. The same is true for big data.

The HR daemon has the ability to let the actors within organizations tinker, and the HR department encourages people to tinker in various ways. More importantly, the process of tinkering is noted as tinkering and, therefore, is potentially baloney. There is a fundamental difference between using big data in the sense of big data baloney detection and for big data tinkering. The difference between tinkering and precise work is comparable to the contrast between bricoleur and engineer (Freeman 2007). Big data tinkering is also about testing the possibilities of technology (Miller, J. H. 2015) and the ways in which big data can be utilized within organizations. Those tinkerers (as the model in World of Warcraft) are crossing borders, be they social or ethical. They are at the very least a higher risk for organizations, so the HR department needs to establish a safe space for such tinkering. In addition to establishing tinker spaces (similar to maker spaces) and encouraging people to tinker, the HR department needs to balance both extremes of rigorous use and wild speculation. One way to deal with this could be through risk evaluation.

Rosetta Mission

A fitting example of such tinkering with big data is the Rosetta Mission (Glassmeier et al. 2007) by the European Space Agency (ESA). In 2004, a probe was launched on a flight to the comet Tschurjumow-Gerassimenko. The mission goal was to land the Lander Philae on the comet. This alone was ambitious, however, the ESA had little information about the comet. In addition, due to the distance between Earth and the comet, steering was impossible (there was an approximately 30-minute delay), so the ESA faced a situation where they had to have everything necessary on board prior to the launch. Just putting everything into the satellite was not an option, and it would have increased the weight to dimensions that would lead to different problems. Every gram not only cost more money, it would also make the launch into orbit more dangerous. A heavier lander would also cause complications in the landing process. So, ESA did not know the composition of the comet and they had to deal with the strong restriction in weight.

Under these circumstances, landing on a comet is a difficult mission, and a great deal of work was required to increase the chances of success. As a public organization, ESA minimizes potential risk or else funding will be spent on projects that are more promising. ESA, thus, ran simulations for a variety of different conditions. Although this is typical of any space flight since the Apollo (Branch 1997), the Rosetta team claimed to be "prepared for every eventuality" (New Scientist 2015). ESA gathered information from various sources, including previous missions, research surveys, simulations, data from other space agencies, and data from suppliers about their components. All this information was used to design Rosetta and Philae adequately. Although information about the comet was scarce, several compositions were nearly impossible, and it seemed plausible that some combination of ice and iron would be realistic. ESA tinkered a plan in which Philae would grapple to the surface with harpoons, an idea that at first sounded quite extraordinary. Above all, the weight question was tackled by several researchers. A variety of simulations were necessary to find a sufficient solution, respecting the interests of every researcher and minimizing the risk of failure.

Although ESA ran a vast number of simulations, the comet proved to be much harder than anticipated (Yuhas 2014). Philae bounced off the surface and eventually crash-landed. Philae came down in a shadowy region and, therefore, could not generate energy through its solar panels. There was little time to gather as much information as possible. The ability to tinker allowed the team to gather much and, most importantly, interesting data from Philae in the short period of time until the battery died (Dorminey 2014). ESA quickly anticipated what was possible and plausible within the remaining time. Ultimately, the Rosetta mission was executed successfully (Lee 2015).

Especially in today's world, organizations need to think outside the box and be creative. Big data enable organizations to think of every eventuality, however, risk cannot be entirely eliminated. Using baloney detection may have led the organization to decide against the Rosetta mission, but big data supplied the tinkerers with enough information to convince the ESA to follow the plan. It was clear that there were risks, but by thinking of all possible eventualities, the team was able to deal with those problems. Philae may not have worked to the fullest capacity, but it gained new insights, and that was the mission.

4.2.3.3 Big Data Risk Governance

The HR daemon faces two extremes, big data baloney detection and big data tinkering. Both are entangled with a certain type of risk and can, as a result, be categorized through risk. The risk value gives top management the ability to make decisions more precisely and be more aware of the surrounding risks. As noted for the core assumptions of big data, organizations are influenced by risk. Especially for the goal of achieving a homeodynamic organization, risks are disruptive factors that could disturb the delicate balance within an organization. In addition to the risks of big data, organizations are still affected by risks from external sources and internal sources. Big data help to make risks visible and transparent. Big data are also a potential risk factor, especially if big data are a black box. Furthermore, globalization and interconnectedness render today's world riskier than ever before. Concepts like risk governance (Stein & Wiedemann 2016) attempt to steer risks in a beneficial way for any organization. Big data and risk governance try to decrease the influence of uncontrollable risks. Neither are currently equipped for an efficient search for risks, or for their precise evaluation, yet both may greatly benefit from one another. I, therefore, propose the unified function of *big data risk governance*.

The research field of risk governance developed in recent years (van Asselt & Renn 2011) and its origins can be linked to the European Commission TRUSTNET program (Amendola 2002). There is, however, no common definition of risk governance. Generally speaking, it deals with the regulation of (commercial) risks (Renn 2008, Stein 2013). In order to define risk governance precisely and understand the underlying framework, a clear look at the terms risk and governance is required.

Risk can be defined as the "effect of uncertainty on objectives" (ISO 31000), and even though this definition describes the situation adequately, it does not sufficiently cover the potentials of risk. An assessment of risk seems feasible, however "understanding these various aspects of uncertainty in a complex system is extremely difficult" (van Asselt & Renn 2011: 437). Risk is also connected to ambiguity (Renn et al. 2011) because risk regulation is always linked to people, and ambiguity refers to the existence of multiple values. This makes risk assessment variable and debatable. Risks can be separated into simple risks, complexity-induced risks, uncertainty-induced risks, and ambiguity-induced risks (IRGC 2005). The first type is rare as risks are rarely simple (de Vries et al. 2011). The majority of risks can be sorted into the other classes, but findings reveal that risks are usually managed as simple risks (van Asselt & Renn 2011).

The term 'governance' refers to several different actors determining decisions, the appropriate framework, and processes (Hagendijk & Irwin 2006). The term 'governance' is derived from the Latin word *gubernare* and the Greek word κυβερνάω. In ancient times, it was connected to the navigation of a boat and the responsibilities of the captain. The term 'governance' within an organization follows the same rationale and describes the process of steering an organization in a certain strategic direction. Governance per se is, therefore, the task of 'navigating through rough waters'. Similar to the definition of risk complexity, uncertainty and ambiguity will influence governance and make the task extensive. As far as governance is concerned, the task is to deal with risks.

Based on the definitions of risk and governance we identify both terms as linked to complexity, uncertainty, and ambiguity. Risk governance is, therefore, a construct that tries to tackle the complexity, uncertainty and ambiguity of risks in a way that is traceable and systematic. Risk governance also includes structures that monitor and give early warning (Charnley & Elliot 2002). Risk governance is not simply a type of risk management; it increases risk resilience (Collingridge 1996). Discussion regarding risk governance has led to a dynamic concept concerning risk. Dealing with risk is no longer an if-then-else loop but a system that is flexible enough to adapt to the prevailing conditions, however, if for risk governance we provide the roles of steerer, captain, and decider, it takes on a superordinate role within organizations.

Big data and risk governance are capable of dealing with risks, however, they are apparently inadequate for dealing with the overwhelming complexity of risks, especially because big data on their own are a source of novel risks. Unifying both functions into one reveals that there are several complementary aspects. Risk governance, on the one hand, requires information in order to search a risk network for potential risks. Lacking accurate information makes steering an organization impossible. Big data support risk governance with an abundance of information (Bell et al. 2009). On the other hand, big data struggle with the evaluation of their objectivity. There is an inherent data bias in any big data analysis. Interestingly, risk governance deals with such shortcomings, and subsequently, such uncertainties and risks on a daily basis. Risk governance is, therefore, capable of supporting big data analysis. Big data and risk governance could significantly benefit from each other. Big data and risk governance enable each other to work more efficiently, particularly in providing rigor and the relevant results for organizations. Based on this, unifying both systems creates one singular function capable of utilizing those dualities.

The function of big data risk governance creates new tasks within organizations. Due to its duality, big data and risk governance cross-fertilize each other. As shown in Figure 10, I propose the following elements: establishing, identifying, seeking, assessing, mitigating, and anticipating. These aspects of big data risk governance enable risk governance through big data and vice versa.

Figure 10: Big Data Risk Governance

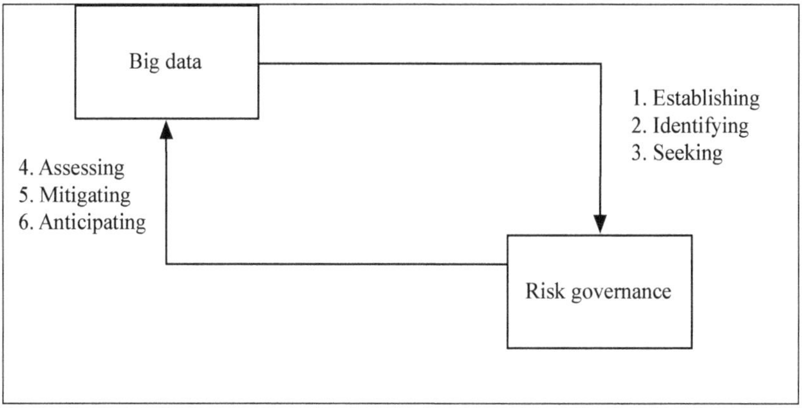

The first element is *establishing*. As stated earlier, the risk network is essential for risk governance. What are the potential risks for an organization? Big data can provide the necessary information for such a task. By analyzing all available data, it is possible to establish a risk network of all risks. That information can include risks that have only a distant effect on organizations, but still are intertwined with them in a small way. On that basis, risk governance obtains a broad but precise picture of the risks surrounding organizations.

In a second step, big data supports risk governance in the *identification* task. Knowing all risks can be overwhelming and can have a paralyzing effect, however not all risks are relevant to an organization. Depending on the risk network, some risks are more influential than others, and on the basis of this information, risk governance can focus on a selection of risks rather than all risks. Big data also provide information about the connections of risks within the risk network. How are those risks connected and how do they interact with each other? Based on the answer to that question, big data can contribute to the *seeking* process of undiscovered risks. Due to the granular picture of the interconnections of risks within the risk network, it is possible to find new risks: in today's complex world in particular, these new risks can be from the result of second-order effects or cannibalization effects (Desai 2001). Although a single risk seems insignificant on its own, in connection with other risks it could be critical. In those first steps, big data support risk governance to get a clearer picture of the risk network and enable risk governance to act in a better and quicker way. This is especially true since those tasks can be done in 'near real-time' (McAfee & Brynjolfsson 2012). Big data can also simulate a variety of compositions of the risk network and develop various predictions.

Big data can also be supported through risk governance. In the fourth step, risk governance improves upon the *assessment* part of big data. As stated earlier, big data are not as objective as some researchers believe (Boyd & Crawford 2012) and that means big data depend on a critical analysis (Dalton & Thatcher 2014). Risk governance can fill this void and provide an assessment of the risk network and the influences of such a risk network. What are the causal relations and do they make sense? Those results need to be comprehended from both a contextualized and a holistic perspective. Due to such thorough analysis, risk governance helps to find errors within the big data analysis and also supports big data in *mitigating* their risks. Risk governance could use several algorithms to minimize the big data risk. Finally, risk governance supports big data in *anticipating* new developments and new risks. Big data, on their own, only find results within their limited data sets. Big data only react to this constructed data world. Every predictive analysis (Sprague 2015) will take place based on that data bias (Scholz 2015b). Risk governance needs to seek new data sources, implement new ideas, and proactively envisage the potential (re) actions of the environment and especially of the human actors within the environment. Reinforcing this effect, humans react to the results of big data analysis, and this could lead to self-fulfilling prophecies (Merton 1948), anticipatory obedience (Lepping 2011), or self-preventing prophecy (Brin 2012). Such behavior will cause distortion, forcing big data to adapt. Big data risk governance can anticipate such behavior as well.

This duality results in a big data risk governance that is capable of acting and reacting in a quicker, broader, deeper, more differentiated, more sustainable, and more insistent way. Due to the dualities of both elements, big data risk governance develops a risk network that helps understand the complex environment in which an organization acts. Big data become more precise, less risky, and are questioned constantly. Big data risk governance establishes an elaborate risk network that needs to be fostered and groomed permanently, and this comes at a high price. As previously stated, there is a dilemma in deciding between big data baloney detection and big data tinkering, however, big data risk governance is capable of revealing its usefulness by showing potential gain and loss and presenting simulated results. It represents an investment for the future. Big data risk governance, combined in this duality, supports itself to overcome its inadequacies. It is reasonable from an evolutionary perspective to combine both worlds into one distinct function, thus giving an organization a function to navigate through the data deluge and through the risk network, leading to the achievement of the goal of a homeodynamic organization in the midst of a turbulent and stormy sea.

On top of these aspects of the big data risk governance, the ethical problem as the ethical component is part of the concept of risk governance (Stein & Wiedemann 2016) and, therefore, will be part of big data risk governance as well. Although, risk governance is a highly ethical topic, big data are the subject of ethical discussions even more frequently. The importance of tackling the ethical question in the field of big data can be highlighted in the following example. In the video game "Starcraft II: Legacy of the Void" (issued by Blizzard in 2015), there is an interesting discussion that highlights the relevance of being critical, and skeptical, and dealing with big data in a social and ethical way.

> "Karax: The replication data is the sort that allows accurate duplication of one's consciousness. Fenix personality may be accurate. Within the ninety-ninth percentile.
>
> Artanis: So there is a chance for discrepancy.
>
> Karax: Quite a miniscule one.
>
> Artanis: And in a lifetime, how many choices does that variation impact? Who would you be with such a difference in the decisions you've made?"

Although copying one conscious soul to another may be science fiction, it can be applied to the big data discussion. As noted earlier, there is a data shadow of people created by big data, which creates a hyperidentity that may or may not coincide with a person's real identity. This example explains that any difference between the data shadow, the social shadow, and the actual identity will have consequences and will influence the actors within any organization. Although there may be only tiny variations between those shadows and identity, over time, these differences could become impactful. As long as big data are not big enough to meet these expectations, big data use will be pivotal. It may be even the case that there will always be some remainder of difference, as long as we are not part of an all-embracing surveillance society. If big data cannot grasp the behavior of people entirely, it will only create a subjective view of the shadow, and analysis depends on the methods used to decrypt such behavior. Big data will only give us a portion of the data shadow and only one shadow from a certain viewpoint, leading to an interpretation that leaves much room for interpretation. As Barry (2011: 8) summarizes it, big data "provide destabilizing amounts of knowledge and information which lack the regulating force of philosophy which ... ensures that institutions remain rational". Organizations deal with a variety of different ethical obstacles in the use of big data. Mittelstadt and Floridi (2015) derived the following ethical themes on the basis of a literature review of 68 papers:

- "Informed consent
- Privacy
- Anonymisation
- Data protection
- Ownership
- Epistemology
- Big data divide" (2015: 10).

Informed consent tackles the question of whether people's consent to allowing data to be collected can become something dynamic in times of big data. People do not know what data are collected about them, or know how such data are used. This leads to privacy and anonymization issues, two themes that are strongly influenced by big data. Privacy is a highly debated topic in terms of big data (e.g. O'Hara & Shadbolt 2008, Solove 2011, Tene & Polonetsky 2012), and there will be several transformations concerning that topic (Rubinstein 2013); anonymization is a concept from the past, and with big data it becomes relatively easy to de-anonymize anonymized data sets. Clemons et al. (2014) call this the myth of anonymization. Data protection is essential for big data within organizations, however, many leaks (e.g. Kuner et al. 2012) reveal that it is still a neglected topic and that is critical as the data sets become more granular and more individualized. Ownership is about the discussion of who owns the data. This is a topic that will be discussed in detail regarding big data authorship. Mittelstadt and Floridi (2015) identified a link towards epistemology. It seems problematic to understand big data and the complexity of big data, in a context where big data more and more mimic a black box. Finally, Mittelstadt and Floridi (2015) deal with the big data divide and tackle it through the divide of power and control over big data. The element of surveillance and profiling is especially highlighted. People are unaware of being profiled or surveilled. All these themes raise questions about justice concerning big data and the difficult task of dealing with big data in an ethical way.

The predominant question for dealing with big data within organizations is, thus, what is ethical big data use. There are two ways in which a moral compass could be derived: on an individual basis or at an institutional (group-level) basis. Both are influenced by each other and focusing on one will harm the other (Mittelstadt & Floridi 2015). We therefore seek alternatives. One way would be to look at a higher order system that connects both ethical perspectives. This could be found in a kind of big data ethos. *Ethos (ἔθος)* is the Greek word for custom or habit and describes guiding beliefs or ideals. An ethos supports its users with guidelines and simple rules to follow. There is still space to act according to individual and institutional ethical elements, but ethos also supplies people with an ethical safety net.

Big data ethos is an omnipresent guiding system that influences the complete big data use process. Ethical considerations are essential in data collection and data analysis. Data shadows and social shadows are competing to influence the perceived identity. Given that there is an inherent bias in big data use, from the viewpoint of a big data ethos there is a need for responsible handling throughout the complete process. The crucial part of big data, therefore, is vigilance. I am proposing a concept of "data vigilance." The term vigilance is derived from the Latin word *vigilantia* and means wakefulness, watchfulness, and attention. Vigilance is important in order to adapt the use to remove any kind of bias. Data vigilance is also linked to someone's accountability.

In order to specify ethical big data vigilance, I propose a framework consisting of four dimensions. *Attention*: This deals with the element of being alert at all times and developing a watchful eye in every situation. *Consciousness*: This refers to the necessity of having some sort of ethical value system or ethos and following its values. *Intention*: The purpose of big data use is to reach certain objectives which go beyond the purpose of maximizing profit and include all interests within an organization, especially those of the employees. *Stabilization*: Analyses on the basis of big data can be done in real time and organizations can be completely flexible, however, the goal is to make an organization stable (not static) within its environment. Big data enable organizations to become more homeodynamic and, therefore, sustainable.

All these dimensions are essential in order for organizations to gain an understanding of what uses of big data may be ethical in their particular case. It is essential to understand that a certain use may be ethical for one organization and deemed unethical for other organizations. On the basis of its organizational signature, an organization already has some insights into a rudimentary version of the ethical value system. Facebook will have a different value system and will, for example, be more open to big data than Airbus. In a way, Facebook's product is big data and, therefore, will focus even more on big data than Airbus. In order to highlight the necessity of vigilance, we can look at the following example. It is possible to insert code into a webpage to retrieve the battery status from a smartphone. The idea of this was initially to deactivate certain functions to save the battery. Although that makes sense at first glance, it is theoretically possible to use this information to identify a particular user on the internet because the information about battery status is incredibly precise (Olejnik et al. 2015). With knowledge about battery status, people can be tracked across the internet and the browsing history can be reconstructed. The HR department is watchful with such information and deals with it in a fitting way.

Big data risk governance is, consequently, helpful for an understanding of the work of the HR department within the HR daemon. Ethical big data vigilance within big data risk governance allows organizations to derive methods that fit, as well as establish structures to proactively find new issues that may be ethically questionable and deal with them. Vigilance will also allow organizations to be more resilient, transparent, and, most importantly, comprehensible to the other actors within organizations. Dealing with big data in such an ethical way will decrease distortion, but also will disenchant the magic of big data. Organizations use big data to improve themselves, and, therefore, treat people within the organization in an ethical way.

Case of Google Flu

Although this example does not completely follow the proposed big data risk governance model, it reveals the relevance of the concept and the need for dealing with the risks of big data and the surrounding risks. In 2008, Google launched a project that helped to predict outbreaks of the flu. Google claimed that their predictions were 97% accurate compared to data from the Center of Disease Control, but without the time delay that CDC results normally have (Ginsberg et al. 2009).

Google used a vast amount of data to *establish* a risk network concerning flu-related searches. They *identified* 1,152 data points that related to the flu (Ginsberg et al. 2009), however, they initially did not *seek* new or abnormal search patterns like the A-H1N1 influenza (Cook et al. 2011, Olson et al. 2013). Those inconsistencies within the risk network caused Google Flu to overestimate flu prevalence, making the results no longer precise, and rendering them even less accurate than those of the CDC (Lazer et al. 2014, Kugler 2016). Those errors within the big data analysis were spiraling out of control and Google needed to *assess* the potential risk sources. As Lazer et al. (2014) note, Google changed the software and the algorithm of their searches. In 2011, for example, they introduced a feature that suggests search terms on the basis of the initial search word. People also change their search behavior (Lazonder et al. 2000) over time and search engines are susceptible to manipulation to a certain degree (Zwitter 2014). Understanding and comprehending those influences is important, but it is crucial to *mitigate* those big data related risks. Lazer et al. present one solution in their paper: "By combining GFT and lagged CDC data [...] we can substantially improve on the performance of GFT or the CDC alone" (Lazer et al. 2014: 1203). There is a high volatility inherent in the internet, and Google is in the midst of all those changes. How do human dynamics interact with algorithm dynamics? It is essential to *anticipate* for future challenges. Google has enough data, but as this case shows, they do not always ask the right questions.

Taking the perspective of ethical big data vigilance, it may seem that many of the problems could be prevented due to having a high *attention*. Nevertheless, some sort of *consciousness* was available, and the *intention* of Google was to improve the health aspects of their users and, therefore, can be seen as positive. As well as the foundation for some *stabilization* in their ethical big data vigilance. But, the main problem seems to be that Google had certain blind spots in their attention. They did not see various risks that led to the distortion of results. Consequently, only the attention was problematic, but this was critical enough to lead to the problems described.

There are many sources that make any endeavor riskier, and big data can contribute to such risk. At the moment, there is a blind spot concerning the risks of big data, although it seems obvious that big data will not always find the correct answers. Big data and the case of Google reveal that it is not a pure, objective, technical entity, but that big data are strongly entangled with the social world and any change will influence their informative value. Algorithms change and people change in such volatile ways that they make big data a risk as well. Therefore, the ethical perspective is crucial within the observation of big data and, furthermore, is an ongoing process. Big data risk governance involves dealing with those risks as well as the ethical consequences and enables organizations to thoroughly evaluate their strategic decisions.

4.2.3.4 Big Data Immersion

In the next step, the HR daemon tackles the integration of big data within the organization. In particular with a focus on the relationship between big data and the people, big data will become immersed in the organization and, therefore, affect many aspects and fields within it. First of all, the HR daemon will need to tackle the questions surrounding data protection, privacy, ownership, and copyright of big data, as well as the people generating the data, which will be conceptualized in big data authorship. Furthermore, big data are not static entities and will change constantly over time and space. Therefore, if big data are an integral part of the organization, it will be necessary to monitor and maintain big data or to deal with *big data curation*. Finally, a main part of the HR daemon will be to train the employees in handling big data on their own and developing essential big data competencies. The HR department will increase *big data literacy* within the organization.

4.2.3.4.1 Big Data Authorship

Using big data within an organization can be beneficial, and is in the bilateral interest of employer and employee, but there are still the main issues of data protection, privacy, ownership, and copyright of big data about employees. It is important to highlight that those terms are not synonyms but rather tackle diverse topics concerning big data (Dix 2016). However, organizations are facing a difficult situation.

On the one hand, complete transparency is not the ultimate goal, as it can lead to information overload (Toffler 1970); on the other hand, hiding all data is also the wrong approach, as it can lead to a violation of trust. It is essential to find solutions to secure data and privacy, as well as to legally ensure copyright. Within an organization, it is essential that some balance is achieved and that both the organization and the employees benefit from big data adequately. That can be difficult, as some use can violate privacy and copyright. As stated earlier, the legal landscape is currently still struggling in relation to big data. One issue is that because of legal regulations data are only used anonymously, but that would cripple big data use at the individual level. Big data can support employees in individualized ways, but if the employees are anonymized this benefit dissipates. In fact, being anonymous is a myth in these times of big data (Clemons et al. 2014); with enough data it is possible to de-anonymize any information (Tene & Polonetsky 2012, Froomkin 2015). Although an HR department does not de-anonymize these data sets, the potential for malpractice is clear. Another issue is that, for example, European law prohibits personalized data use if a specific purpose is not given, and the tools of big data such as data mining are legally highly restricted. Following the law to the letter would mean that exploring big data is not allowed within organizations in any way.

Although data protection laws are more rigid in Europe than in any other part of the world, problems with big data are not limited to European organizations. Keeping employees in the dark and abusing big data leads to the post-panoptical (Bauman 2000) behavior of employees. Because employees believe they are monitored, they change their behavior appropriately in the sense of anticipatory obedience (Lepping 2011), for example, "cleaning up" their Facebook profiles (Brown & Vaughn 2011), and thereby distorting the data shadow and the social shadow even further. Other employees could discover the patterns of surveillance and exploit the system (Zarsky 2008, O'Neil 2012), as well as making the shadow of their identity vaguer. People changing or hiding their behavior will lead to an impreciseness of big data and subsequently to errors in decisions based on *infected* data.

The question of what type of privacy is even possible in today's organizations also arises, in organizations, in which movements are traceable, sensors are ubiquitous and smart machines are collecting a massive pile of data. It becomes increasingly difficult not to gather data about employees, not even information from secondary sources. Smart machines depend on their sensors and for security reasons need to keep track of the humans around them. This information about the people could potentially be repurposed for different objectives. Within an organization, people are constantly tracked, deliberately and unwittingly, thus making their data shadows bigger (not necessarily more precise) and contributing to the hyperidentity of employees within organizations. The question of rights regarding data is even more unclear in that case. Do we assume that the person (or organization or even machine) that collects the data has all rights on the data, or that the person the data are about holds the property rights?

Privacy (Matzner 2014) and copyright laws (Kaisler et al. 2013) are apparently unfit to deal with such modern problems (Lessig 2008), and, therefore, I propose a concept of *big data authorship*. The idea is rooted in similar observations in virtual worlds (Roncallo-Dow et al. 2013). In those worlds, "authorship is a collaborative act" (Guertin 2012: 2) and goes beyond the question of copyright and privacy. In this case, both the player within the virtual world and the creator of the virtual world create the design and story of this virtual world together. They are both its authors. Although privacy and copyright are still difficult to grasp, both parties understand and see their task as producing and contributing to a common goal, and in some cases, these interactions evolve or emerge in a form of unwritten social contract and mutual trust (Roncallo-Dow et al. 2013). Virtual worlds such as 'World of Warcraft' and 'Eve Online' are built on these premises of collaboration and both games have now existed for over ten years. People are becoming more committed to remaining loyal to a game if they perceive the authorship to be fair and truthful.

In the context of an organization, gathering big data is also a collaborative act to which everybody within an organization contributes. Due to the complexity of data, it is difficult to untangle these contributions. If we understand big data within organizations as a similar concept to the authorship of virtual worlds, then big data are a shared experience and joint action between organization and employee. Big data are first of all kept within organizations, and the HR department will act as "primary gatekeeper" (Grimes 2006: 970). Keeping the data generated by an organization and its actors within the organization will increase the trust of employees. They will more freely share their information if they know that the data are safe and secure.

Everybody is seen as the author of big data within organizations, and the HR department is responsible for the fair use of big data. HR departments can flag certain data as private or as having limited visibility, and employees can do the same. If employees are interested, they can use existing data for their own analyses following the motto: putting big data into the hands of employees. The HR department fosters this relationship and monitors fairness within the organization. A social contract, as in the example above, may be a broad solution, however the HR department could also use the tool of psychological contracts (Rousseau & Tijoriwala 1998). Everybody collaboratively contributes to big data within organizations and is seen as the author of that. The HR department needs to implement the ability to have data transparency as a part of the HR daemon, allowing employees to use the data (and tinker with big data). The HR department also needs to evaluate the appropriateness of hiding certain data, avoiding a potential transparency trap (Bernstein 2014).

Big data are not limited to employees within organizations, but include employees who have left organizations, and who have authored a variety of data in their time within organizations. It would be possible for the HR department to define a certain data set involving these employees and cleanse it from internal data. Such personal data sets can describe several performance indicators and serve as a datafied certificate of employment. In analogy to encryption, this data set could be compared to a personal key. These data are cleansed from critical information about the organization but are meaningful for the employee. Furthermore, this key can be combined with the organizational signature (Wang et al. 2014, Stein et al. 2016), and would generate a simulated assessment of the employee within any organization.

Such personal information could be an interesting addition to the recruiting cycle. First of all, employees would have a reason to share their data willingly because it benefits them, and, secondly, any organization could check the fit of a potential employee more precisely. It is important to emphasize that only with a combination of an organizational key, on the basis of the organizational signature and a personal key, meaningful and contextual results can be derived. The organization or the top management will trust the HR department to be responsible for those personal keys, and especially for the organizational key.

Although legal discussions concerning copyright and privacy are ongoing, big data authorship is a proactive task for the HR department which utilizes big data in a more transparent way. It reduces the power imbalance and enables an employee to contribute to the use of big data. Trust will be strengthened, and the employees will include some (encrypted) transparency, as well as keeping the useful part of their data. Big data are, at least within organizations, not something blurry that somehow emerges out of nothing, but something to which everybody is actively contributing. This change from being constantly monitored without knowing it to the idea that such data can be used by both organizations and employees to improve the organizational environment and an employee's career, marks a strong psychological change in narration. From a legal perspective issues are not solved by authorship, but it enables organizations to find a solution that allows them to utilize big data for their own needs. The emphasis now is more on perceived fair use and the trust relationship between organizations and the employees. This is something HRM was designed for, and only the HR department is capable of dealing with these interrelationships within organizations.

Quantified Self

In light of current digitization and the technological progress in wearable devices, people are now capable of self-tracking (McFedries 2013). The process of self-digitization, self-monitoring, or self-quantifying is combined in the term 'quantified self'. People are willing to share their data in order to be compared to other users and to evaluate their performance. Quantified self-movement is, currently, often discussed in the context of health topics (e.g. Swan 2013, Ruckenstein & Pantzar 2015) and the motivation can be described as follows:

> "... technological developments in the portability, precision and 'accuracy' of heart rate meters has transformed the realm of everyday calculability. They allow us to 'see' our own heart (instant feedback), and in seeing, allow us to make adjustments in what we do: they allow us to quite literally tune our own engine" (Pantzar & Shove 2005: 5).

There is the ideal that such self-tracking will be about one person all alone, so the claim is that n = 1 (Nafus & Sherman 2014). However, the step linking such data with big data is relatively small, and in the context of organizations there are ways to track people's communication. With the help of badges, it is possible to track and analyze the communications of all employees within an organization (Orbach et al. 2015, Atzmueller et al. 2016): who talks with whom, for how long, at which location and the topic. This can be linked with analysis of the voice: Is the person agitated? Combining the quantified self-data with communications and voice data would give many insights into employees.

It is obvious that such data need to remain within organizations, and it is also obvious that the decision-maker within organizations will have some knowledge of the people involved. If, as in the paper by Orbach et al. (2015), the goal is to improve informal communication within organizations, somebody needs to know which people are being analyzed. This information may not be of relevance in the results of the analysis, but there will be somebody within the organization that had access to such data. Using data from wearables would become even more personal. Data will also be available from smart machines, for example, infrared sensors could unwillingly monitor employees, and these data may be useful to organizations.

Data about employees would become more personal and more detailed than ever before. Big data within organizations are unwillingly full of data shadows which are not anonymized in any way. It becomes essential to have somebody who watches over the employees and allows them to use their data as well. Organizations and employees author big data within organizations and, in order to utilize such data to the maximum, people need to trust and believe in the fair use of such data. Interestingly, quantified self-movement shows that people are willing to contribute their data if there is an actual incentive, such as better health (Swan 2013), and a certain trust that their data are protected (Nafus & Sherman 2014).

As in any organization within the self-tracking business, big data within organizations depend on the people and their self-interest to contribute their data. There is a need for trust of the data fiduciary. Any tracking and, therefore, big data within an organization is surveillance, but the task of the HR department is to make the experience convenient for everybody involved (Whitson 2013).

4.2.3.4.2 Big Data Curation

In the company context, it is essential to remember and to note the missing objectivity (Gitelman 2013) and interference in the way data are gathered, analyzed, and interpreted (Van Dijck & Poell 2013). False claims of objectivity will have an impact and will disrupt the relationship between employer and employee. Making the big data value chain transparent within organizations and additionally incorporating the subjective bias into the analysis will improve the relationship. Both sides will be increasingly able to understand and discuss the results. The HR department acts as a moderator for such communication, a task that is already present within organizations.

Although organizations have a variety of data available and generate more data all the time, as described in big data risk governance, big data are not jacks-of-all-trades. Depending on the way big data are analyzed, there are different sources of risk. There is a general risk, but there are more specific risks from big data such as subjective interpretation, contextualized data, statistical biases, sampling biases and so on (McNeely & Hahm 2014). There is, thus, a *big data risk additive* that the HR department incorporates into big data analyses. Due to such an additive, it seems that big data and the results of big data are subjective, and need much work to be transformed into results that can be used by an organization. Mayer-Schönberger and Cukier (2013) envisioned new professionals called algorithmitists, who "would act as reviewers of big-data analyses and predictions [...]. They would evaluate the selection of data sources, the choice of analytical and predictive tools, including algorithms and models, and the interpretation of results" (2013: 180). I argue against the idea that such professionals understand, at first, the inner life of organizations and therefore have knowledge about the organizational signature and also the competencies for the analysis of big data. A technical expert would not be suitable, but a social expert could deal with the inadequacies of big data within organizations.

Big data are often contextualized, subjective, and consist of repurposed data. The amount of data solely collected about people and for the purpose of HRM is relatively small and often categorized as bad data (Buckingham 2015). Many processes within organizations are outsourced, automated, or robo-sourced (Gore 2013). Data are generated that do not follow certain standards within organizations or are not available for further use. Big data also deteriorate over time and become less precise and riskier to use. There is a similarity with the half-life of knowledge, in which the knowledge of people and their competencies may become obsolete over time, but the period of time varies from knowledge to knowledge. Big data will become obsolete over time, yet the speed at which data dissolves depends on the data. The big data half-life adds to the risk additive.

The HR department curates big data within organizations and deals with existing big data and the acquisition of big data. The task is similar to that of a museum curator. Big data need to be presented in a certain form to fit a distinct theme, and such a theme could be the organizational signature. Big data need to be checked and refurbished if necessary. Big data from other sources that may be useful are controlled and adjusted to the organizational signature. The origin of any data is however clearly labeled and the changes made to the data tracked. If it is unknown when and where certain data were collected, there is a high risk that such data will be highly subjective and highly outdated. If the data collected from a reliable source are fundamentally changed and distorted by that in certain way, such data are no longer reliable.

Data are always interlinked with other data within big data. Changes in one part of big data influence other parts of big data. If the HR department discovers errors, they need to correct them at this certain point, but also need to check all links connecting to the error. The curation process has, consequently, the potential to be self-healing for an organization, with a focus on big data. Errors are also bound to happen, especially in a turbulent environment and with heterogeneous data sources, and if the data are contextualized in an inept way. Somebody has the ability to control and curate data in a distinct way that fits the organization, and to archive data that no longer seem required or seem outdated. Alternatively, in contrast to the half-life of knowledge and the need to unlearn such knowledge, big data can file such outdated information, put it in an archive and access it at any time. Only the required data and the current data are on display.

A variety of *control and test* mechanisms are necessary. Finding errors or distortions within big data are a critical task for the HR department and the ability to do this is implemented in the HR daemon; however, as big data are vast – at the organizational level, individual level, and relational level – it will be not sufficient to control only one level. It is necessary to have the ability to combine levels so as to spot problems. For example, a combination of distant reading (Moretti 2013) and ground truth (Pickles 1995) will be essential in order to triangulate the effects and identify consequences on organizations. Distant reading is an approach to understanding "literature not by studying particular texts, but by aggregating and analyzing massive amounts of data" (Schulz 2011), a description that fits any big data approach. Accompanying such a distant picture is a method from the field of cartography, where researchers use data from the ground to support their analyses. A similar metaphor can be used with big data analysis. Although a large amount of data allows a picture from *far above*, it is also essential to validate it from the ground. The ground can mean the individual level, but it also can mean the methodological inner life of an algorithm, so the HR department can look into the heart of their big data analyses. Especially in times of machine learning, those algorithms act on their own in a certain way. Using the analogy of the museum's curator, they will not want museum pieces to be categorized without knowing how the categorization works. The curator can try to make sense afterwards and reverse engineer the algorithm behind it, but if the algorithm sorted the pieces inadequately it will take time and resources to rearrange the museum pieces. The same is true for big data: letting the algorithm do the work may sound promising at first, and if it works it works, but there is a risk that the big data within the organization will be transformed into something irreversible.

The HR department implementing the HR daemon, therefore, needs to monitor, handle the risks, collect, curate as well as control, and test big data within the organization. It also needs to detect and categorize data shadows and social shadows, biases and track changes. The organizational signature is the masterpiece of the collection and is treated and preserved in that way. That task may not be done exclusively by HRM, but they lead the curation, and are responsible and accountable for big data curation within organizations.

Employment Screening

As noted earlier, it has become customary to do background checks on employees through big data. Often this is done by checking the activities of the potential employee on social media. It is much debated (e.g. Sorgdrager 2004), whether these results are appropriate for categorizing potential employees. Social media profiling (Esposti 2014) is part of employment screening, and "there truly has been an explosion in how technology has changed and continues to change selection practice" (Ryan & Ployhart 2013: 20.11). Not only do organizations use social media, but they do extensive background checks. Organizations use a variety of sources and in that way use external vendors of information.

Another way of screening employees in the U.S. is by evaluating their credit scores, as provided by one of three scoring companies, and this is used in many organizations (Bernerth et al. 2012). A recent study by the SHRM (2010) discovered that 43% of organizations (n = 385) checked their job applicants on the basis of their credit score if they are potentially selected, and 13% of organizations run a credit check on all job candidates. A credit score describes the creditworthiness of a person. Organizations use this score to make assumptions about potential employees (Hollinger & Adams 2008) and predict their behavior and performance (Gallagher 2006). Normally, organizations would not obtain numerical values but rather information about how much money is owed to whom (Kuhn 2013). It seems misleading to use such broken down values as there is a variety of information lost in obtaining the score. Reasons that could explain a low credit score or poor credit report are not available, such as race, residence, or family status (Traub 2013). That in itself is problematic, and leads to a new type of financial discrimination (Shepard 2013).

Although the implications of credit checks are questionable, organizations rely on the data delivered by those external scoring agencies, and they depend on the accuracy of those credit scores. Choosing an employee on the basis of a credit score and realizing afterwards that there are errors in the credit score could have an impact on the selection of the best candidate. Eventually, the best candidate may not be identified as the best. The credit score is not as accurate as some people believe. The Federal Trade Commission (FTC) ran a survey in 2012 and discovered that "26% of the 1,001 participants in the study identified at least one potentially material error" (FTC 2012: i), and even worse, 5.2% had an error in their credit score that would lead to a lower interest rate for a credit. Consequently, a credit score may or may not be accurate and in a follow-up study the FTC revealed that people who disputed their credit score had a "meaningful credit score increase" (FTC 2015: ii). A credit score will have an impact on recruitment if it is incorrect, and by a percentage of 26% there is a high probability that there will be errors within the credit score of most job candidates.

The task of big data curation, in this case, is to incorporate the potential risk into the calculation. A credit score can probably give some insight into the history of any potential employee, however, there is a margin of error. The credit score needs to be flagged as a subjective score, and the credit report as a subjective source of information. At the very least, any potential employee will have the chance to give a comment on this information. Errors may be unknown to them or there may be other reasons for this score.

The element of discrimination especially is seen as critical. Recruitment on the basis of data or numbers could disguise discrimination behind a veil of objectivity. As Traub (2013) found, there is a discriminatory factor within the credit score. This factor can be discussed in regard to the Chinese Credit Score or social credit system (Stanley 2015). The Chinese government is truly applying big data to a universal score. Due to the high regulation of the internet, they can collect a massive amount of information about people and, additionally, connect a person's information to their friends. The score is then evaluated not only from an individual's behavior within society, but from how their friends behave (Falkvinge 2015). The score determines whether people can apply for a visa or a loan (Hua, 2015), and is seen as "the most staggering, publicly announced, scaled use of big data" (Obbema et al. 2015). Although the rating is at the moment available to the state, people brag about their scores on social media (Doctorow 2015) and the next step is to use this score for recruitment, especially as it is more granular, and made on the basis of more information than the American credit score. This sounds like science fiction, but as Doctorow summarizes it:

> "Paternalism, surveillance, social control, guilt by association, paternalistic application of behavioral economics and ideology-driven shunning and isolation – it's like someone took all my novels and blended them together, and turned them into policy (with Chinese characteristics)" (2015).

Organizations cannot change such systems, but they will factor in all the problems such a system would mean for an organization. Although such scores are seemingly accurate, they are not. There may be a social agenda behind them, but big data are predominately subjective, erroneous or simply outdated. Such errors are difficult to eradicate (Pasquale 2015) and so the HR department deals with the risk additive. Big data are in dire need of curation in a form that means organizations can use them in an efficient way. Blindly introducing big data into organizations will change the organization in an uncontrollable direction. As for a museum curator, however, there are ways to transform big data to fit with organizations.

4.2.3.4.3 Big Data Literacy

The role of people within organizations is currently transforming fundamentally. Machines and computers are becoming the grunt workers for many narrow and repeatable tasks. This development is also observable in conjunction with big data:

employees gain room to focus on complex thinking, innovation, and creativity. This depends, however, on the utilization of big data. At the moment there is a disparity between people who have the ability to use those new technologies and people who are not able to do so extensively. The former are augmented by technology and capable of doing incredible things, the latter, however, fall into a veil of ignorance and are driven by big data to a certain degree. To make matters worse, at the moment there is a war for big data talent (Ahalt & Kelly 2013), in which government agencies and IT companies are competing with every other organization. Organizations need to close the big data gap and recruit or train potential candidates. The HR department needs to improve big data literacy within organizations (Christozov & Toleva-Stoimenova 2015). D'Ignazio and Bhargava describe the concept big data literacy as follows:

- "Identifying when and where data is being collected
- Understanding the algorithmic manipulations
- Weighing the real and potential ethical impacts" (2015: 2).

Talking about big data literacy reveals the connection with media literacy: "Media literacy – indeed literacy more generally – is the ability to access, analyze, evaluate, and create messages in a variety of forms" (Livingstone 2004: 5). People are enabled to deal with media, and such a description fits big data literacy. Big data literacy is about these competences, and the task of big data literacy is to train employees in such a way that they are capable of dealing with big data. The HR department has the capacity to encourage their development. By means of human resource development, people can be taught big data literacy. This training will tackle computational thinking (Wing 2006), statistical thinking (Hoerl & Snee 2012), and skeptical thinking (Sagan 1996). The goal is to empower the employee to open the black box and lift the curtain behind the big data magic. As Clarke (1977: 35) stated in one of his three laws, there is the tendency to understand such complex and opaque technology as magic: "Any sufficiently advanced technology is indistinguishable from magic". Big data contribute to the veil of ignorance (Rawls 1971) within organizations. In order to be able to deal with the task, the HR department needs extensive training in computational thinking (data farm), statistical and skeptical thinking (fog of big data), and utilizing their HRM and ethical training (big data watchdog).

The prime goal of HR development is to lift this veil of ignorance so that employees understand the use of big data within organizations. Employees also need to be trained in a way that means they are also capable of tinkering with the existing data, and exploring on their own. Achieving this goal will be done through training and development in big data competences. This also includes the HR department, and as Priestly stated precisely, "We're all data geeks now" (2015: 29). It will be essential to lift all employees to a level that they understand and use big data analytics (Davenport 2013) as well as being critical of them (Boyd & Crawford 2012).

Depending on the big data literacy within organizations, there will be a tendency towards convergence or divergence, and standardization or individualization. Employees need to be capable of dealing with big data. John Draper, aka Captain Crunch, coined the term "Woz-Principle" (Freiberger & Swaine 1999), derived from an idea by Steve Wozniak. It suggests that as many people as possible are trained in using technology to the extent that they are capable of inventing new things. At best, technology is as simple and open as possible. From this it follows that people are empowered to design their working environment for their specific needs (Baumgärtel 2015). Such trends can be seen in the open source communities, hackers, or in gaming. By empowering employees in the sense of the Woz-Principle, the HR department will transform the operation system or the HR daemon of the organization and individualize the working environment or user interface for every employee. Employees will be able to customize their working environment for their specific needs. This could lead to a realization of the following statement: "Making people think is the best that a machine can achieve" (Gigerenzer 2015: 320). That means that the goal of HRM is to enable people to have an intrinsic "desire to exploit the information capacity of the new technology" (Zuboff 1988: 392).

A critical issue is that people tend to have a certain amount of technophobia (Brosnan 2002), anxiety (Beckers & Schmidt 2001), and a fear of coding (Spinellis 2001). Although it may sound promising to follow the Woz-Principle in training and development, and beneficial to training the data geeks on their own, the focus lies in convincing people to learn to code and to use statistics. The HR department is the leader in the role of transforming organizations to enable people to design their own tools. It is responsible for a balance of user-friendliness and for the ability to tinker. The essential task is to convince people to acquire the basic abilities of computational and statistical thinking (Dasgupta & Resnick 2014). Empowering people through the Woz-Principle will let them think, create, and innovate in a way that leads to a prolonged competitive edge for any organization.

This means that all employees need a rudimentary training in computational, statistical, and skeptical thinking. Big data influence all decisions within organizations and employees will face big data on a daily basis, but big data are complex by definition, so organizations are transparent concerning their analyses; employees who do not understand the consequences will be skeptical and deny the use of big data (Shah et al. 2012). Ignorance may be bliss, but only with improvements in big data literacy the effectiveness of big data can be improved for organizations and for every employee.

There is also a new layer of complexity concerning learning and development. Today's big data algorithms are no longer mere tools (Varian 2014); they are learning as well. Machine learning (Goldberg & Holland 1988) and deep learning (Deng & Yu 2014) are standard parts within algorithms. Algorithms learn on their own and, most importantly, change on their own (Gillespie 2012) – and, if not watched, become unintelligible to humans (LaFrance 2015). This means there is a dependency, or even duality, between human learning and machine learning. Human learning and machine learning are also within a feedback loop, and are (negatively speaking) in a vicious cycle or (positively speaking) a co-evolutionary loop. This is similar to the idea of the red-queen hypothesis (van Valen 1973), in which both sides challenge each other to improve, adapt, and learn. The function of developing and training is no longer limited to people, but includes algorithms as well. This is especially important since algorithms can learn erroneous things (just as humans can), but algorithms are not capable of judging them. Algorithms can, therefore, develop ideologies (Mager 2012) and subsequently create reality. HR development and machine learning will merge into one function within organizations in the future. People are trained and algorithms are trained. Both are constantly working together, so they influence and learn from each other.

The HR department is the expert in training and development, they are also capable of dealing with resistance to change (Dass & Parker 1999) and convincing people (Armenakis et al. 1993). It becomes increasingly important to train the employees in big data literacy, not only to achieve some form of transparency, but also to harness the possibilities of big data. "Data is useless without the skills to analyze it" (Harris 2012). Big data will unfold all capacities if people are taking advantage of the potential. The borders between HR development and machine learning are also dissolving and training for algorithms as well as for employees will help employees to work better with big data.

Gamification of HR development

There is a current observable trend not only of *gamifying* work (Oprescu et al. 2014) but also of gamifying HRM, by, for example, incorporating video game design elements into HRM processes. Gamification (Hamari et al. 2014) is often used under the premise that gaming is fun and engaging. Players trying to win a game are highly motivated to reach high scores. This is of particular interest for managers, which makes it understandable for HRM to jump on the bandwagon of gamification. There are several definitions of gamification: "the process of game-thinking and game mechanics to engage users and solve problems" (Zichermann & Cunninham 2011: XIV) or "gamification refers to: a process of enhancing a service with affordances for gameful experiences in order to support users' overall value creation" (Huotari & Hamari 2012: 19). The most commonly cited definition reads: "gamification is the use of game design elements in non-game contexts" (Deterding et al. 2011: 1).

It could be interesting to gamify HR development, especially as learning curves are an integral part of any game (Rosser et al. 2007), however, using a gamification system 'off the shelf' will be a source of irritation (Bogost 2014) and will lead to resistance (Deterding 2014). Big data may help to make the system fit organizations, and the gamification system will stay static and finite (Nicholson 2012). Interestingly, video game developers already utilize massive amounts of data to understand their players and adapt their games to their player base. People have different interests and different skills, subsequently, this diverse player base will influence the way they are playing the game. Video game designers act on the knowledge they acquire and design the most fitting experience for these players, so they stay within the game and play it. A game, and massive multiplayer online games (MMOs) depend on a dynamic development of the world to keep their players bound to a particular game.

Big data within video games like 'World of Warcraft' help to individualize the experience of any player and keep the player within the flow (Csikszentmihalyi 2010). The learning curve can especially be individualized. Players learn new elements of the game at their individual speed, so they are not overburdened or bored. The bar is constantly raised (Scholz 2015c). The game conveys a sense of mastery (Nicholson 2012) and enables players to narrate their own story.

Such a concept can be transferred to HR development for big data literacy and the HR department can implement and cultivate such a gamification system within the HR daemon. People will learn about big data literacy in a playful way, and, thus, lower their big data phobia. They will learn at their individual speed, but will learn to become better equipped to deal with big data. Such a system can be designed in a similar way to the tools within the HR daemon and train employees to program their own tools, following the Woz-Principle.

4.2.4 Human Resource Centaur

The HR department reacted to the transformation towards a homeodynamic organization through big data with a new role and created the HR daemon. Both actions require us to deal with big data and the impact of big data on an organization. However, both are still more reactive than proactive. In chapter 2.4 it became evident that the reaction to big data can be polarizing, however, it seems that an augmentation of both worlds would be beneficial for the organization. Big data will not lead to a competitive advantage, but the people augmented by big data will be the source of competitive advantage. Consequently, it will be essential to enable the workforce to exploit big data and augment them by using big data. Until now, big data have changed the role of the HR department and the way it works. In addition, while big data are now everywhere in organizations and immersed completely into them, big data are somewhere in the background and are something that seems to have no direct connection to the employees.

The goal is, therefore, a way to put big data into the hands of the employees. The HR department's task is to design a frontend, in which the employees can interact easily with the HR daemon and the available big data within organizations. The goal is to give the employee a "'cockpit' interface on their computers that they help design" (McDonald 2011). The idea is similar to the concept of augmentation described by Davenport and Kirby (2015) as "starting with what humans do today and figuring out how that work could be deepened rather than diminished by a greater use of machines". Augmenting people with big data depends on the system that is implemented, and this frontend system I will conceptualize under the term *HR centaur*.

Why a centaur? Looking at the evolution of chess, it is well known that Deep Blue beat Kasparov in 1997. Today the best players follow the concept of centaur chess. Human and machine team up and augment each other in an extraordinary way, superior to human and machine alone. "Centaur chess is all about amplifying human performance" (Cassidy 2014). Such a collaboration of human and machine, as observed in chess, has proven to be far superior to playing alone (Ford 2015). Humans can focus on their creative and innovational roles, delegating the grunt work, or at least the operative tasks, to big data tools. Big data can aid and will help "human beings think smarter" (Kelly 2014). Collaboration, "if we handle it wisely, [...] can bring immense benefits" (McCorduck 2015: 51).

In a popular song by Daft Punk, the band sings about "work it harder, make it better, do it faster, makes us stronger", and this metaphor is strikingly fitting to the modern world that is enhanced by big data. Big data enable organizations to gain access to an abundance of data and use them for their purposes, however, most organizations drown in the glut of data (Emerson & Kane 2013), and are surrounded by an opaque data fog. It is, therefore, one of the most important tasks of HRM to deal with big data in an efficient way and build a sustainable infrastructure. Gaining a competitive edge or even a competitive advantage out of big data use is a more strategic challenge. People are augmented by big data. They can *work it harder*, as they can specialize in their competencies and use their capabilities efficiently. They can *make it better*: they have a different point of view and so see problems and obstacles that the other would miss. They can *do it faster*: dynamics and velocity are crucial for the success of modern organizations. Working together, division of labor is more precise and, synergies are used in a more fitting way. This *makes us stronger*: such an organization is more capable of tackling a situation in its environment. It can adapt to new challenges and govern the risks surrounding them. Tinkering and performing with virtuosity will lead to the essential competitive edge any organization needs. The HR centaur needs to act as a multipurpose tool kit (Zuboff 2014), to enable people, especially as:

> "[M]achine intelligence does not lower the threshold for human skills – it raises the threshold. Whether it's programmed financial products or military drones, complex systems increase the need for critical reasoning and strategic oversight" (Zuboff 2014).

The HR centaur will augment employees to be able to deal with the increased threshold and give them all the essential tools to exploit big data to a potential competitive advantage. One way to implement such an HR centaur system is to reevaluate the potential of gamification and learn from video games. Big data are, per se, digital, so the link between big data and people is digital as well. Gamification and video games are normally embedded in the digital realm and, therefore, there are many ways to learn from those digital pioneers. I have already defined gamification in a previous chapter. Although gamification can contribute to the HR centaur and there is the potential to increase transparency, individualization, and strategic agility (Stein & Scholz 2016), such a system would be predominantly designed by the HR department. They will act as gamification designers (Raftopoulos 2015), and will constantly update the system to fit the needs of organizations. Apps will be built that will make employees transparent (Buchhorn 2015), however, the HR daemon and the engine behind these will stay shrouded. There are many ways to analyze employees and give them information back about their work. Various components of a game can be translated exactly towards such HR centaur software (Scholz 2013c). For example, a talent tree can show an employee what things are available to learn and what specific programs fit the current job. People can be matched as teams on the basis of their ELO scores (Erhardt 2016), a method to calculate the skill level and rate people on the basis of the score, and so on, but these systems are one-directional from the HR department to the employee. It seems that big data will act as a bridge between video games and the real world; for example, metrics and indicators used in video games are more and more available in the real world due to big data. Hocquet (2016) described this bridge on the case of Football manager video games and the increasing entanglement to the football world and the datafication of the football world. The challenge will be to create a HR centaur system that will be designed by the HR department and the employee, thereby following the Woz Principle in the truest sense.

Radical Gamification

But what would such a system look like? In a conference paper, Stein and Scholz (2016) envisioned a concept of a radical gamification. The following example is derived and adapted to the context of this thesis from the conference paper.

Contrary to most gamification within HRM, which can be characterized as *casual* gamification, a *professional* gamification with proper design, intensive planning and careful coordination could tackle existing problems in a new manner. In order to avoid "gamification is bullshit" critics (Bogost 2014) and the reproach of engaging in pure 'gamewashing' (in analogy to "greenwashing" in the corporate social responsibility debate (Dahl 2010)), I will present a short example of a proposed HR centaur system.

Radical gamification of HRM would best be possible in an organization with a non-existent or underdeveloped HRM function (although that does not mean no HR department) and a workforce open to change. They are also able to gather big data and have a rudimentary understanding of the HR daemon. An organization of that kind would be a start-up, mainly consisting of "digital natives" (Prensky 2001) with basic programming competences, and in a field that allows them to gather data digitally: an IT startup. Cultural elements of gamer culture (Shaw 2010) and hacker culture (Levy 2001) would be beneficial. Only under those conditions would employees be intrinsically motivated to participate and to increase the *gamification rate*. They have also the ability to utilize big data in an efficient way and, therefore, the organization will have an interest in using big data in such a way.

The radical gamification of HRM starts with the basic idea that all the HRM functionalities that are needed by employees or by management are to be developed bottom-up. Everybody is entitled to write add-ons such as apps and to modify those emerging functional worlds (Sotamaa 2010) as needed. The *look and feel* of that gamified HRM, then, imitates the design of a sandbox game like 'Minecraft', where the players can do whatever they imagine: a holiday scheduling system, performance measurement, monitoring presence, multi-project management, a team task assignment support tool – the possibilities are endless. Incentives can be coupled with gamification contributions. No longer being a traditional HRM department, a HR gamification designer will be given the task of supervising the gamification system and simply acting as a corrective. Everything else will follow the market principle and the logic of self-organization. An employee in need of a specific functionality simply *buys* an existing tool from the market or programs it autonomously. The lack of a distinct functionality can be understood as a strong indicator that there is simply no need for it in the organization.

Such radical gamification is scalable and develops concomitantly with an organization's growth. The integration of the employees who need to acquire competences in utilizing and developing *their own* HR centaur system is crucial. Issues such as relative fairness will be tackled so that people cannot cheat or exploit the system. It interferes, but never with self-organized teamwork or competition. Nor does it cancel them out.

In a system of that kind, employees shape their range of HRM and at the same time live a gamification culture to its fullest. The HR gamification designer merely acts in the background supervising the people-related *engine* or the HR daemon of the organization. Stenros (2015) talks about second-order design backed by Salen and Zimmermann: "As a game designer, you can never directly design play. You can only design the rules that give rise to it" (2004: 168). Leveraging transparency, individualization and strategic agility will benefit employees and the organization – both mutually increasing their ability to make homeodynamics work.

4.2.5 Big Data Membrane

At this point the homeodynamic organization is fully implemented due to the integration of big data within the organization. The HR department has reacted with new roles, created the HR daemon, and proactively augmented its employees through the HR centaur. But these changes also lead to a high transparency concerning big data and potential critical information about the organization. If everybody has access to most of the data within the organization, keeping the data within the organization will be complicated. I postulated that big data plus people will generate a competitive advantage, but a competitive advantage which is only possible if this knowledge is kept a secret. The problem is that big data are truly everywhere. Big data seem to be unbounded and free floating. It is, therefore, essential to have ways to protect personal data from the outside world, such as personal data about employees, data about the organizational signature, and data that describe the competitive edge of an organization.

In nature, there is a semipermeable membrane that selectively allows an exchange between the outside and the inside. Such a big data membrane would be capable of deciding what data are shareable and what data are critical and, therefore, will not be shared under any circumstances. Some data can be exchanged freely; others will be kept within the organization. In the context of an organization, this is comparable to open innovation. Chesbrough (2006) explicitly points out that the only innovations which are shared are those not critical for the competitive advantage of an organization.

One way to deal with data sharing is to focus on the membrane and improve the selection of such a membrane, however that implies a critical reflection of what is valuable and what is not. Big data are known for being vulnerable (Newman 2015b) and it may be beneficial to keep critical data in one place rather than outsource them to the cloud (e.g. Kraska 2013). Big data, thus, need to be encrypted and people need to be trained to follow the encryption rules. Both elements can be achieved through the HR department, however, every encryption is breakable. In a recent paper, Zyskind et al. (2015) linked the protection of big data to the concept of a block chain.

> "A block chain is a type of database that takes a number of records and puts them in a block (rather like collating them on to a single sheet of paper). Each block is then 'chained' to the next block, using a cryptographic signature. This allows block chains to be used like a ledger, which can be shared and corroborated by anyone with the appropriate permissions" (Government Office for Science 2016: 17).

In addition to achieving a certain transparency and traceability of big data, which is beneficial for big data within organizations as well, the data are protected in a relatively strong way (Swan 2015). Organizations will be able to encrypt their organizational signatures and will have a ledger of all changes. The ledger decreases the risk of manipulation from inside and outside. Employees know that their personal data are encrypted and their personnel file is completely transparent to those people that are allowed to see it. A block chain can be used to secure it from the outside and make it less susceptible to manipulation. It is, furthermore, a way to identify changes within the big data. Additionally, it enables organizations to have a form of time machine and retract changes (interesting for the data farm) as well as reconstruct after manipulations. A block chain would not be able to prevent corporate espionage, however, it would just make it more difficult for anybody to steal information. Although the concept may sound a bit futuristic at the moment, block chains will influence organizations and radically transform them. Tapscott and Tapscott are using the words "agility, openness, and consensus" (2016: 90) as well as "decentralization" (2016: 91). They are, furthermore, talking about the importance of the code of the block chain system that seems comparable to the organizational signature.

Another issue involves the information that is sent through the membrane to the outside environment. In the common practice of competitive intelligence (Kahaner 1997), for example, it becomes increasingly less difficult for organizations to gather all the information available about their competitors. This information makes organizations more transparent from the outside, though organizations can proactively work against it. Big data enable organizations to gather external information as well and can reconstruct the picture competitors have of them. An organization can then play proactively and spread either correct information or false information, altering and distorting the picture the competitor has. It is, thus, possible to improve the protection of certain knowledge that is linked to competitive advantage, by flooding big data into the outside world in order to mislead the competitors.

Both acts are enabled through technology and big data, but it is essential that they are not driven by big data. Big data are shackled by the computational logic and, therefore, susceptible to being de-encrypted. This computational logic of big data needs to be transformed into an irrational protection system and the people within organizations must be capable of supplying irrationality. These critical data can be easily translated into a gamified system. People can be sent on missions to hack the system from within (in a secure environment) or sent outside to spread misinformation. Due to the possibility of obtaining a *precise* picture of organizations from the outside, employees will see the effect of their work and be intrinsically motivated to keep certain elements a secret and to spread other information.

Apple Car

There is a hypothetical case about an Apple car by Shen et al. (2011), where they discuss a potential extension into the automobile sector. Today, there is still no official information regarding an Apple car, but many people (e.g. Harris 2015, Hotten 2015, Jones 2015) seem to know that Apple is working on an automobile. We are, therefore, talking about a hypothetical case, but it reveals the existence of a big data membrane around Apple.

There are many data on the Apple car and many rumors on the car, however, it seems that Apple is, in a way, directing the information. They are using this leaked news to get something out of it. Take another example: it seemed that everybody knew that Apple was working on a television. Apple never talked about the project, but it got a clear picture of the chances of such a device on the market. Without ever openly talking about the project it realized that there is potentially no profitable market for such a product. In the case of the Apple car, CEO Cook made it clear that they would not comment on the Apple Car, but at the same time he teases people about it:

> "Yeah, I'm probably not going to do that. The great thing about being here is we're curious people. We explore technologies, and we explore products.
>
> And we're always thinking about ways that Apple can make great products that people love, that help them in some way. And we don't go into very many categories, as you know. We edit very much. We talk about a lot of things and do fewer. We debate many things and do a lot fewer" (Cook in Lashinsky, 2016).

In times of big data, information is ubiquitous and organizations cannot control all information concerning their organization. Information is leaked and rumors are spread, however, those rumors can be steered and governed. Media companies have focused on Apple concerning the Apple Car project and harvested a great deal of big data to gather new information. Apple is not capable of producing a car on its own, and therefore have an army of subcontractors and suppliers. What technologies hide behind Apple's current products (e.g. batteries)? Apple monitors information about the Apple car and sees where the information originates. The media is currently monitoring Apple's recruitment efforts, leading to a list of potential project members (Kahn 2015).

This is all especially important, as Apple is always mysterious about its new products. Although it is still unclear whether there is a car in development, Apple can focus on developing a car and everybody else can speculate about the chances of this car on the market. This is relatively cheap market analysis. They get to know what customers want and what they dislike. Even if they put the car in mothballs, they have probably improved the battery technology of their laptops, tablets, and smartphones.

Big data increase the risk of losing a potential competitive advantage and make organizations more transparent than ever before, although it seems that organizations can steer the data stream to a certain degree. If organizations invest resources in the big data membrane, they will be capable of exploiting this apparent weakness. They can convert it into a strength by utilizing the abilities of big data and people combined to the fullest, so in a certain sense, we are talking, in analogy to centaur chess, of *centaur intelligence.*

4.3 Homeodynamic Goldilocks Zone

The complete implementation of big data within any organization through the new roles of the HR department, HR daemon, and HR centaur, will enable the homeodynamic organization to be more dynamic and consequently, capable of gravitating around a certain form of balance. Big data are always about a strategical decision between polarities, as shown in the core dimensionalities mentioned earlier, and the positioning between those polarities. Organizations constructed in a dynamic way will be able to correct their course, and though as they are complex systems, small changes may have big impacts and oversteering due to time-lag is always a possibility (Liu et al. 2011, Diesner 2015). This is true especially as big data within an organization will make the organization potentially faster, but real-time remains an illusion (Buhl et al. 2013). A homeodynamic organization will probably not be able to achieve perfect homeodynamic balance, but will keep organizations close, especially as it is not necessary to balance everything out exactly.

Organizations need to be in the right zone, that is the so-called 'Goldilocks zone'. The term is derived from the story of 'Goldilocks and the Three Bears', in which a girl searches for things that are "just right" (Spier 2011: 148). The term emerged and gained popularity for describing the zone of solar systems in which planets are potentially habitable (Kasting et al. 1993). The concept proposes that planets need to range in a certain zone of variables to be habitable that may depend on distance from the sun, luminosity of the sun, size of the planet, certain elements (e.g. helium) available in certain amounts and so on (Lineweaver et al. 2004). A homeodynamic organization will also be stable within a certain *homeodynamic Goldilocks zone* as shown in Table 15.

Table 15: Positioning of the Homeodynamic Goldilocks Zone

Polarities of the Core Dimensionalities in a Homeodynamic Organization		
Data Linearity	Homeodynamic Goldilocks Zone	Data Monodology
Data Rigor		Data Swiftness
Data Island		Data Assemblage
Social Constructivism		Data Constructivism
Data Risk Avoidance		Data Risk Seeking
Social Shadow		Data Shadow
Self-Determined		Data-Determined
Data Reliance		Data Bias

Organizations need to find a way to deal with changes and evaluate influences on their position within the zone. With the HR daemon, the HR department, and the HR centaur, organizations are capable of keeping themselves 'just right', however, there are several constraints that will be incorporated into the calculations. These are (1) the organizational signature, (2) the trust climate, and (3) the rate of dynamization and the complexity parameter. The organizational signature is the core DNA of any organization, therefore it will not be changed all the time. Consequently, the organizational signature seems to be a fixed influence on homeodynamic organization. The trust climate is critical and influences the reaction time of an organization. Without trust in the HR department or in big data use, there will be resistance and distrust. Changes will not be implemented and organizations will drift into a lock-in situation, and depending on the current situation, potentially move outside the homeodynamic Goldilocks zone and "fall apart completely" (Spier 2011: 148).

The next constraint is the degree of dynamization and complexity. People tend to prefer a static, orderly, observable, and linear environment, but reality resembles the opposite (Maguire et al. 2011). In order to categorize the facets of dynamization, Stein (2015: 3–4) presented the following:

- More dynamic in the strategy-related sense of 'more differentiated'
- More dynamic in the mechanics-related sense of 'faster'
- More dynamic in the organics-related sense of 'more versatile'
- More dynamic in the culture-related sense of 'more strategically agile'
- More dynamic in the intelligence-related sense of 'more methodologically competent'
- More dynamic in the virtuality-related sense of 'more flexible'

Although it makes sense to improve the dynamization, there is tradeoff in the sense of complexity. Homeodynamic organizations are complex systems and big data increase the complexity furthermore. The elements of unpredictability, non-equilibrium and non-linearity in particular (Maguire et al. 2011) lead to unexpected threats, opaque and secondary effects, and uncertainties within a complex system (Dörner 1989). Organizations will deal with those potential risks.

In a nutshell, it is possible to keep organizations within the homeodynamic Goldilocks zone, however, it is a complex task that is supported and disturbed by big data. Depending on big data use, and, therefore, depending on the HR department, organizations gain the ability to remain within the zone. It sounds like a difficult task for any organization and a costly project to transform an organization into a data-augmented homeodynamic organization, but it will increase the survivability of any organization drastically. People will become *the* competitive advantage of organizations and they will transform big data into something more than just the standardized tools many organizations are currently using: although it is expensive, utilizing big data in this extensive way allows the management to have a precise view of their organization. People are no longer an opaque cost pool, but their contribution can be accounted for. At the very least, utilizing big data will be beneficial for all employees, as it allows everybody to focus on their strategic and innovational input, and delegate the operational and automatable tasks to the HR daemon.

5. Results

5.1 Summary

Big data are transforming the world (Mayer-Schönberger & Cukier 2013) and organizations will deal with this transformational wave. It is leveling the playing field and everybody has access to nearly all information today. With the current developments of 3D printing and automatization, organizations are facing an even more comprehensive change that may shake the foundations of the construct corporation. Coase (1937) was puzzled by the question about the nature of the firm and he tried to explain it by transaction costs. There is an inherent reason to organize within a firm and that is if an organization has lower costs than the costs on the market (under the assumption of an imperfect market). Although the argument is still valid, technological progress changed the reason for the existence of firms quite a bit.

Information is freely available and organizations are endangered by the ease of being copycatted (Hota et al. 2015): One file in the wrong hand is enough and the competitive advantage of some technical gain is gone. In the future of 3D printing, everybody could produce everything, and suing everybody for their intellectual property is not the solution. An example is the fight of music companies against file-sharing (John 2014). Companies like Apple or Spotify realized that people did not want to pirate music, but have easy way to access it (Richardson 2014), and that furthermore they were willing to pay for the music. Organizations still exist as a result of the human factor (Zuboff 2014). Using a music example again, in the wake of the piracy discussion, many musicians engaged with their customers in a more direct way. They improved their social media performance and focused more on live performances. In the abundance of today's music, those musicians stand out in a crowd. Coase (1937) talked about the benefits of organizing resources within a firm and that is still the truth, however, the focus is on binding people, both employees and customers. Employees make organizations unique, so customers stay loyal to the organizations. Employees are transforming big data into a competitive advantage. In times where everybody can produce everything, the human touch will be the difference, and keeping that spark within organizations will be a strong reason that these are still firms in the future.

A general question we need to discuss is: Why and how will an organization tackle big data? Using big data will be mandatory for any organization in order to survive. Davenport (2014) states that any industry within the following categories will be reshaped: an industry that "moves things, sells to consumers, employs machinery, sells or use content, provides service, has physical facilities or involves money" (Davenport 2014: 33). It seems that any industry falls under one of the categories and we already see that any industry can benefit from big data. It is, therefore, crucial that any organization specifies its big data strategy and does not let it remain an unclear technology (Cohen et al. 1972). Not using big data will lead to a significant competitive disadvantage: if there is useful information available, the competitors will use it. In the Moneyball example, the Oakland Athletics had the advantage as innovators (first mover advantage) for one season, but afterwards, everybody used their statistical approach. In order to keep up and not fall behind, organizations will use big data and will implement big data structures. By becoming standardized, those organizations will keep up, but will generate a competitive advantage only by accident. Big data need to be introduced in organizations to support and augment the people. In this way, organizations will be fundamentally transformed into homeodynamic organizations.

Big data will have an impact at the social, organizational and individual levels. Big data are everywhere (George et al. 2014) and big data influence everything. Although big data are rooted in the digital world, the current developments reveal that the digital world and the real world are moving towards each other, and will merge more and more. There is however, a dangerous assumption here. Big data are objective, and therefore far superior than people's subjective gut-feeling. There are various reasons for the subjectivity of big data. Big data are gathered from various sources and so are embedded in a certain context. Even though big data are big, they are not big enough and will never depict a complete picture of everything. There are blind spots, there are errors, there is subjectivity (be it introduced by people or algorithms), and, consequently, big data may not be used autonomously for data-driven decisions in any way. Big data are also personal, and any organization may view big data that way. Outsourcing may seem profitable, but will lead to a standardization of organizations. It is important to take into account that although the way that big data are solving our problems sounds magical (Reeves et al. 2015), putting big data into a black box is a new source of risk (Pasquale 2015). One error in an algorithm doing flash trading (Buchanan 2015) led to a disruption in the Dow Jones Index that day. The market got off fairly lightly, the organization behind the error was snapped up cheaply. The error was discovered quickly but big data and especially artificial intelligence are developing as fast as they work, and we *do not* understand them (LaFrance 2015, Adams in Byrnes 2016).

Big data are ubiquitous and influence everything, but are opaque, distort reality, and will be biased. Many companies that utilize big data are claiming that the subjectivity of people is harming them, and big data are less biased. Recruiting can be used as an example. C. C. Miller (2015) complains that hiring decisions are often biased and many potential candidates are dismissed for personal reasons. The tendency to overconfidence increases as people rely on data for hiring, but the algorithm could be discriminatory, so there is a paradox at hand. Although big data are able to find more potential employees, in the end these selections will rely on some bias predicted by correlations within big data. If we look at Silicon Valley, it becomes abundantly clear that there is a diversity problem (Pittinsky 2016), however the data may reveal the following correlation: young white male engineers are the best candidates, as the majority of engineers are young white males. Although the correlation sounds weird, it will probably be highly significant. Furthermore, every person employed on the basis of that correlation will increase the significance level. Over time, the correlation will influence the structure of organizations and shackle them in a certain direction. Big data roaming freely through organizations will not be beneficial and, therefore, people will deal with it.

Big data act like a force, depending on their use and how people are utilizing big data, however, big data, as the name implies, are huge, and therefore depend on intensive change within organizations. I have proposed the concept of homeodynamic organization which deals with the influence of big data within organizations and enables organizations to harness the positive force of big data. On the basis of that assumption, I conceptualized new tasks and consequently new structures within organizations, namely the HR daemon and the HR centaur, implying that the HR department would be most capable of utilizing big data in order to achieve a competitive advantage. But why will the HR department be accountable for the HR daemon and HR centaur? Both concepts sound relatively technical. The IT department would be capable based on its knowledge, but they are the executors of big data. Although the technical element will still rely heavily on the IT department, and it will stay that way, it would be difficult to handle the implementation and deal with the social and ethical impact on organizations as well. There is a clash of interests, as big data within organizations will not always be implemented in the most feasible way. Giving employees access to big data may sound like an interesting idea, so as to enable employees to be innovative, but from a cybersecurity perspective it sounds like a nightmare. There is a reasonable tendency for the IT department to minimize the freedom of employees and standardize their applications (Sahay 2003). The accounting or finance department would mostly focus on the expense and would have difficulties balancing the interests of employees and the company. Using big data in the proposed way may be a gamble, as organizations wager that their employees might generate a competitive advantage through big data, so that organizations invest resources in order to customize big data use in the most fitting way. When many companies are trying to release organizations from the burden of dealing with big data, it may be not favorable to let these departments be accountable. The organizational department would be a fitting department; however, in recent years they have become cannibalized in companies by IT, HRM, and general management, so that they are now seldom part of a typical organizational structure (e.g. Thom 2006, Stein 2010a). The legal department lacks the ability to cover the emotional elements of leadership that are linked with sensitive ethical questions. Following the legal rule at the moment would also lead to an over-regulated system that does not enable tinkering with big data at all. Finally, the company's top strategic management would also be enabled, however due to the magnitude of changes and their constant interaction with big data it is questionable whether they have the time to deal with it. The HR department therefore seems to be the most fitting department, especially as the focus is on people and not on big data. Big data are transformed in a certain way so that people can generate the innovations needed for achieving a competitive edge.

By appointing the HR department as the integral driver of big data and consequently the enabler for the homeodynamic organization, the importance of the HR department is strengthened. It may be the last chance the HR department gets to take a strategic position within organizations. Current trends (e.g. Cappelli 2015) reveal a loss of importance, however, if people are the competitive advantage in the future the HR department will step up its game. Interestingly, the tasks and functions envisioned in homeodynamic organization are not novel for a HR department. Big data enable organizations to uncover hidden things and get a more concise view of fuzzy elements, and therefore, big data can act as a catalyst for a professionalization (Stein 2010b). Interestingly, the gamification introduced in the HR centaur is also considered an activator for professionalization (Stein & Scholz 2016). If technology is no longer the differentiation factor on the market but people are, networks will be increasingly important. Big data are capable of discovering informal networks and utilizing them, be this formal or informal, and it can be beneficial as well. There is an even stronger interest in supporting emergent hidden networks and disguising shadow networks (Stein et al. 2016). The HR department is capable of monitoring the use of big data so that employees can utilize them quickly. It can generate trust and uphold it, train employees adequately, and exploit big data in the mutual interest of organization and employees.

The HR daemon is probably the most alien feature of the data-augmented homeodynamic organization, however, it is critical to minimize the potential distortion of big data. It also adds scalability capacities within organizations. Big data may be vast, but dealing with big data is often relatively narrow. It seems sometimes that there is an underlying assumption that every big data set can be dealt with by the same algorithm, however, if algorithms are creating their own reality, organizations can get stuck with a distinct data-constructed reality which may or may not be fitting. This may or may not be baloney (Sagan 1996) and, consequently harmful for organizations and the people within them. It is also possible that data collection will be neglected and organizations depend on one certain data supplier, assuming the context of this certain data supplier. There is the tendency to destroy the variety of big data because it is already too big, however every step leads to a certain distortion and increases the power of a construction of reality through data constructivism. The data farm sounds highly technical and the HR department will need the support of the IT department, however from a strategic perspective, it is essential to establish acceptance within organizations for such data farms, and that is a typical HR task. The fog of big data is similarly dependent on the knowledge of statisticians and the risk department, however, balancing both extremes is again a strategic dimension that needs to be discussed with the relevant people. Finally, the big data watchdog is the natural task of any HR department, although supported by the legal department and in extensive corporation with the works council (Hoeren 2014). The HR department, therefore, needs to learn several new skills and roles, but they are already experts in the most critical skills: enabling people to commit to organizations to their fullest potential.

The interface between big data and the work of employees will be interesting in the future (e.g. Chan 2015). The digitization of work will increase even further in the next years. Wearables and sensors will make the employee fully transparent, and although the HR daemon generates trust, all this data is not a one-way road. There are ways to use big data to support employees, especially through tools like augmented reality. The example of gamification is not that far-fetched. HR developments in combination with big data and wearable technology (Park & Jayaraman 2003) sound promising. People can be trained and learn from their mistakes. The information from training can be used to improve it, by focusing on areas where people are struggling. Apps on the computer, smartphones, and wearables (although employees can produce them on their own) will be supplied by the HR department. There are already examples where the shift schedule is processed automatically and incorporates personal constraints. The HR centaur can be designed to be real-time and the HR department to act as a gamification designer. Many elements used in video games sound extremely close to their counter-parts in HRM (Scholz 2013b, Stein & Scholz 2016). Programming those tools is apparently less difficult and it is a habit of players in video games to alter their game (Roncallo-Dow et al. 2013) and customize their user interfaces (Taylor 2009). Somebody however supplies the platform, monitors the fairness and maintains the HR centaur.

This leads to the new role of HRM within organizations. As already stated, big data are a chance to strengthen the position of the HR department and make up lost ground. Big data from a strategic perspective is not about big data in general, but the interaction of big data with employees within a homeodynamic organization, and the new role at a meta-level, is quite similar to the new roles of Ulrich et al. (2013). They are specified slightly differently to those roles by Ulrich et al. (2013), but the general idea is similar. It may be that these new tasks are a bit outside the comfort zone of HRM, but they are essential for reaching the point of utilizing big data. Many of the new roles reveal a certain tendency towards a technology affinity, and in order to fulfill this new role the HR department dismantles the HR-IT barrier and actually achieves a certain form of integration. On the basis of the proposed roles, the HR department will not evolve into big data enthusiasts but uphold the importance of the human factor. It is undeniable that everything is getting more technical, and the HR department develops "awareness of the choices they face, a desire to exploit the information capacity of the new technology, and a commitment to fundamental change in the landscape of authority if a comprehensive informating strategy is to succeed" (Zuboff 1988: 392).

All of these elements contribute to a homeodynamic organization, a theoretical construct that seems only achievable through big data and the humane utilization thereof. Such an organization will be more dynamic and will constantly change. Despite being a loosely coupled and free-floating resource, big data also show that such organizations will gravitate and fluctuate around some form of consensus at the edge of both chaos and order. As the Dartmouth survey revealed (Wang et al. 2014), there is something unique within any organization (Stein et al. 2016). This organizational signature will be influenced by the organizational culture, identity, history, and most importantly by the people within the organization. They often choose to apply to a certain organization based on the organization's appearance and will perform self-selection. Employees will stay and commit to organizations if there is an overlap between individual identity (not the hyperidentity or stage identity) and the organizational signature. Big data unearth the organizational signature in a relatively precise way and thus, contribute to the understanding of organizations. Having knowledge about organizations is not necessarily new, but it becomes more precise through big data. The signature is also something that will not change drastically over time and organizations, in order to stay in the homeodynamic Goldilocks zone, will gravitate around the organizational signature.

Big data will have an impact on organizations and the HR department. There are many ways to deal with big data, but ignoring big data will be fatal. Big data will lead to many discussions from a social, ethical, and legal perspective, however, most emerging out of the fear that people, in this case users, customers, and employees, do not know what is happening (e.g. Dwork & Mulligan 2013, Aradau & Blanke 2015). Making the employee part of big data use will not make those problems go away, but taking the actors within organizations on board will create a solution for this distinct organization. In the interest of surviving on the market, organizations will not try to cheat and prey on their employees, but will want to find meaningful solutions. Big data will decrease the potential to contribute to change and competitive advantage over time, but big data will increase the importance of the employees within an organization to make the essential difference.

5.2 Limitations

Big data are vast and big data are never objective. Big data will create a certain data-constructivist reality or enable a social-constructivist reality through social programming. Those are general assumptions derived from my research and utilized in this theoretical experiment in imagination, however, there are currently around 167,000 hits for big data (May 2016) on Google Scholar and those are only the hits for the exact use of the term 'big data'. There are many other sources (e.g. books, conferences, blog posts, business studies, and newspapers) that contribute to the discussion. Google finds 57 million hits for big data. My research is mainly influenced by the data-constructivist influence of search algorithms and by the social-constructivist influence of my social environment.

This distortion in big data knowledge and my stance towards big data may explain my confidence about the impact of big data and their ability to transform the way we work, although there are many who believe the same (e.g. McAfee & Brynjolfsson 2012, Davenport 2014, Mayer-Schönberger & Cukier 2014). My confidence in the HR department being able to answer the challenge, however, is shared by only few other researchers (e.g. Pentland 2010, Huselid 2015). The general idea of looking at the potentials of big data within organizations follows the proposition of J. H. Miller (2015) to *think* of all possibilities of a new technology before substantiating the social and ethical borders in society. Consequently, I expanded my research into science fiction literature, which can help as "reminders about scientific fact" (Dourish & Bell 2013).

Big data are already transforming society and the governments, IT companies, and marketing are especially driving the big data change. They are admittedly doing a great job in putting big data into a black box. It is understandable that the NSA or Target will not share their algorithms with society. The reaction of society is also understandable; people believe that Big Brother and Little Brother are watching them everywhere and all the time. Big data has a reputation problem, especially as these social and ethical borders are still vague or non-existent. A major problem is that big data apparently have a life of their own, or at least in the perception of society. On the one hand, resistance to big data increases, but on the other hand, powerlessness increases as well. It will, thus, be questionable whether any organization is able to implement such free and transparent use of big data. The HR department deals with the general mistrust of big data and the task of convincing people that big data are not used to surveil or control them.

There is a general question about using technology these days. The proposed form of organization relies heavily on the relevant knowledge concerning big data. Basic knowledge about statistical analysis is essential in order to critically reflect on big data analyses (Dalton & Thatcher 2014). The trend to question the boundaries is also beneficial for the process of knowledge discovery, however understanding complex systems is difficult to teach and to learn. Due to the current information overload that people are already facing, there may be the tendency to *not* deal with big data extensively in the sense I envisioned here.

The concept of homeodynamic organization is heavily rooted in organization theory, and therefore there is a chance that it is an unrealistic, theory-driven vision. This is a problem that all underlying theories need to tackle. Cybernetics, systems theory, population ecology theory, and complex systems theory all describe organizations in a highly abstract way. Homeodynamic organization has the same limitation. Furthermore, emergent behavior and self-organization are an integral part of those theories. These organizations are not controllable in the narrow sense, and big data will not magically make an organization steerable. Organizational inertia will remain. The question that arises is whether the benefits of becoming a homeodynamic organization infused by big data, and data-augmenting employees, really outweigh the costs. The HR department may be able to answer the question, but they need to use big data to prove the efficiency of big data. They trained the people within organizations to be skeptical and to question big data. This sounds like a difficult quandary.

Big data are currently perceived as something that will lead to a surveillance state in the sense of Huxley or Orwell, or to some form of utopia. There is apparently nothing between those extremes. The proposed application of big data within organizations tries to highlight the relevance of people and the potential of big data to augment this. It is, however, a rather positivistic view on big data. The technology of big data has evolved quicker than the reaction of society to big data. The scientific community in particular struggles with social and ethical solutions to big data usage (e.g. Barabási 2013). This thesis attempts to provide a solution within organizations, however, it relies heavily on a transparent, honest, reliable, and ethical way of utilizing big data. This may be possible within an organization, and maybe only possible within organizations of certain sizes, but thereby is creating potential solutions for the general public. There is a highly contextualized solution for dealing with big data in a certain organizational environment, but it works there and nowhere else.

Another element I deliberately neglected in the thesis is the legal implication of big data. The topic is complicated and elements like privacy or copyright will be challenged by big data. There are ways to implement such legal requirements, but they demand a tradeoff. Personal data that can only be used for a distinct purpose will contradict big data, but this is currently the EU legislation. It will be interesting to see if there are other solutions within organizations. These solutions could mimic the psychological contract (Rousseau & Tijoriwala 1998) and include a fair use agreement between organization and employee, however in order to generate the essential trust, organizations are obliged to be in compliance with the law. Consequently, the legal perspective will be a restriction for big data within organizations and the definition of an adequate solution will be a task for labor law researchers.

There are several limitations in this thesis, but the limitations also highlight the essential need for an organization to deal with big data at all and most importantly on their own. Big data are influencing society and the effect will grow with the digitization of society. People and organizations have growing data shadows and these data shadows can be changed in a similar way to a stage identity. It seems that the analogy to big data as a force is quite fitting. There is a majority that is influenced by the force but cannot deal with it. Some feel the force and only a few can use the force for the light side or the dark side. Although this is a strong limitation, it may give better understanding of big data altogether.

5.3 Implications for Human Resource Management

Derose (2013) reports that the HRM professionals of Google have stickers with the mantra: "I have charts and graphs to back me up. So f*** off." It is quite a strange slogan and it may be a sign of data-driven HR, but the sticker may throw people off track, as the goal is to "complement human decision makers, not replace them" (Setty in Derose 2013). The analyses Google are performing are apparently rigorous, and they want to help "HR get a seat at the table" (Jackson 2014). Many things are still unknown about Google's use of data and the company is not known for its transparency, however it has talked about its Project Oxygen and stated that a good leader is a good coach or someone who has a clear vision and strategy (the complete list can be found here: http://goo.gl/CPXIpR), and those results are apparently unimpressive. Since then, Google has launched the platform re:Work and shared at least some of its knowledge, however considering the organizational signature of Google it is reasonable that its HR department will need data to convince people in organizations. Interestingly, the company realized that big data would augment the work of HRM and that HRM decides:

> "Google was founded by, and is still dominated by, engineers. So as it started to hire thousands of people and needed to think more deeply about management over the last few years, it took an intensely data-centric approach. But not everything can be distilled down to an algorithm: even for Google's engineers, automation has its limits" (Nisen 2014).

This challenge is sometimes described as a trap. Luca et al. (2016: 97) claim that "algorithms need managers, too". They claim that algorithms are literal and are black boxes, and consequently, "the challenge for us is to understand their risks and limitations and, through effective management, unlock their remarkable potential" (Luca et al. 2016: 101). However, what will an organization actually do? Big data within an organization are a big challenge but it has become necessary for any organization to deal with big data. The current state of the HR department for example is dire:

> "Number-crunching is easier said than done. Some human-resources departments lack the statistical talent to design and run, say, a multivariate regression analysis, which examines numerous data streams. A number of enterprise software vendors say they provide tools to automate data collection and calculations, but human intelligence is usually necessary to tease out which data are meaningful" (Silvermann 2012).

Although, it seems obvious that HRM will play a key role in the implementation of big data in organizations, the HR department will face a fundamental transformation of its perceived role. However, the first implication is to make the HR department capable of understanding and dealing with big data. Consequently, it is critical to close the big data gap in the HR department. It will not be sufficient to hire a bunch of data scientists (Davenport & Patil 2012), as these people often lack the HRM background. Therefore, the HR department requires big data literacy, consisting of computational thinking, statistical thinking, and skeptical thinking alongside the required business skills, human resource skills, and the knowledge of the organizational context.

Any technological change will trigger some sort of resistance. As discussed in chapter 2.4 this resistance will also happen in the HR department. Big data implementation will lead to an emotional discussion. Consequently, there may be people who resist such a change and need to deal with this resistance (Ford et al. 2008) and to achieve acceptance towards such a change (Sagie et al. 1990). The HR department needs to deal with this resistance and to create acceptance towards big data (Drumm & Scholz 1988). It is essential that the HR department is on board with the organizational changes provoked by big data. The people of the HR department will act as promoters of change and change agents (Caldwell 2001). Bladt and Filbin (2014) categorized four types of employees based on their attitude to big data: data enemy, data skeptic, "data?", and data friend. Using this terminology, the HR department has to be a data friend, in order to persuade the rest of the organization.

Furthermore, big data could lead to an increase in the HR-IT barrier and detach the HR department even more from the rest of the organization. However, big data are only at first sight a technological phenomenon, actually, big data are a social phenomenon. Understanding big data as a technological driver may lead to a retreat of the HRM from the big data problem and handing over the responsibility to the IT department. However, the example of Google already revealed that this is not a path accepted by the employees. The HR department thinks about its role in context of big data. But, it could be that it is forced to step up its game. Due to digitization, it becomes evident that the task of HRM will be highly entangled with big data and consequently that neglecting big data will intensify the basic necessity of HRM in organizations altogether (Cappelli 2015, Charan et al. 2015, Stone et al. 2015). Dealing with big data will increase the survivability of the HR department and, more importantly, transform the big data topic into a social topic.

Big data will transform the way organizations are built and how people work. Big data are a people topic. Smart factories will not work without big data, however, they are already driven by data. However, the biggest transformation is within the people and their place in modern organizations. Many jobs will be transformed by digitization and require new skills. Today there are more secretaries than ten years ago, however, the skill set has changed drastically over time (Bessen 2015). Big data will increase the velocity of change. Big data will, therefore, also increase the velocity of HRM and add up to the various reasons for becoming more dynamic (Stein 2012). The impact of big data is far more comprehensive concerning the people in any organization and, therefore, HRM needs to make big data a people's topic; HRM has to be part of any decision made about big data implementation and, even more, of decisions derived from big data.

Only if the HR department is convinced of the usefulness of big data and is part of the big data implementation, the big data strategy as well as established required structures, can it promote big data within the organization. People will be skeptical about any big data in the organization and the changes accompanying big data. The HR department will be a capable promoter of change and will have convincing arguments for these changes. It is convinced of the changes, it shares responsibility, it understands big data, it recognizes the impact of big data on the employees, and it has the competences to persuade the employees. Furthermore, from all departments involved in such change management, the HR department will be the most fitting department for leading such change. The HR department will have overcome many obstacles concerning employees in the process of the big data implementation. Other departments like the IT department will not see certain obstacles as they have a different orientation to tackle the task of big data implementation; but the HR department is behavior-oriented rather than algorithm-oriented (Scholz 1984) and consequently focuses on the employees.

Finally, big data will augment the work of HRM as well as increase the complexity, especially if the HR department accepts its duty in terms of big data and its impact on the organization and the people. The HR department will transform its role and requires new competencies, however, big data and digitization will release the HR department from pure operative and simple tasks and let it focus on strategic tasks. Therefore, the importance of the HR department will increase as well as the impact of its decisions. Consequently, the HR department moves towards a more professional approach in terms of expertise, differentiation, continuity, and governance (Stein 2010b). In terms of expertise, the HR department requires relevant and up-to-date knowledge about big data and current trends of big data in HRM. Differentiation means that the HR department will follow a differentiated approach depending on the context and situation. Furthermore, it also means a differentiated view on big data and the influence of big data. Though continuity seems to be difficult in terms of big data and the move towards homeodynamic organization, the HR department will focus on continuity or keeping the organization within the homeodynamic goldilocks. Aspects like the organizational signature will stay stable over time. So, continuity will be seen as a differentiated topic. Finally, the aspect of governance highlights the relevance of the HR department in terms of big data. The HR department achieves visibility in the topic of big data in the organization and assumes responsibility for big data within the organization and the creation of homeodynamic organization.

5.4 Implications for Research

It becomes clear that big data require a big theory (West 2013, Monroe et al. 2014, Boellstorff 2015) and rely on various theories (Mayer-Schönberger 2014). Big data will not herald the propagated "end of theory" (Anderson 2008), but quite the opposite. Theories are required to grasp the implications of big data in all multifaceted elements. Everybody will be influenced and, depending on their scientific context, big data may impact them differently. Big data, thus, need a big, overarching meta-theory and additionally many field-specific theories. The goal is to spark a discourse about approaching big data from a theoretical perspective. It would be fatal, however, to conduct big data research without any theoretical framing. It is a rather empiricist view of big data that counterfeits a general objectivity (Silver 2012), but big data are subjective, big data will be influenced by the subjectivity of people, and big data will influence this subjectivity. There is data-constructed reality, social-constructed reality, and a mix of both realities. Without understanding big data theoretically and with the lenses of the respective fields, big data will distort reality in a way that researchers cannot comprehend.

However, from a technological perspective big data are a self-runner, especially with the current evolution of algorithms, artificial intelligence, and data processing power. It seems that either the sky is the limit, or there will be a dystopian fear of the Skynet from Terminator, in which the computers rise up against humanity. Quite the opposite case applies to the social perspective. In the course of this thesis it became evident that big data are a social phenomenon and, therefore, have an extensive impact on society, organizations, and individuals. But without an overarching umbrella theory, big data will have vast implications on basic research as well as applied research.

Basic research will be influenced by big data; many research streams analyze the fit between their theories and big data. In chapter 2.2 it was shown that big data can be linked at the socio-technological level and that big data intensify the reciprocal relationship between big data and society. It is no longer possible to analyze technological phenomena from a deterministic point of view. Consequently, the research stream of science and technology studies will deal with a blurring border between society and technology. Society influences big data and big data influence society at the same time. New and old topics gain momentum. Big data can discriminate. Outliers are neglected and there is a tendency to the mean. The definition of the mean, or what is the standard, is not that objective and can be influenced in various ways. Although it is clear what average means, the question of what the average is becomes unclear. There are many ways big data are not portraying the real situation and are unsocial and unethical, although it will be debatable what social and ethical means and such discourse will be critical. Currently, however, the social and ethical elements are under-researched, even though the impact of big data on people and society is clearly visible. Qiu (2015: 1089) describes the situation between technological progress and the current ethical lag as follows: "Just because it is accessible does not make it ethical."

Another research stream tackled in this thesis is the research in organizational theory. Many theories in that field predate big data and in this thesis cybernetics, systems theory, and population ecology theory were analyzed; furthermore, the unifying theory of complex systems, although this theory is vaster and so more extensive. All these theories as well as the entirety of other organizational theories (for an overview see Kühl 2015) do not deal with big data. I presented several links between some theories and big data, yet it is to be debated if these conclusions are describing the relationship between organizational theory and big data precisely. The effects of big data on organizations are still unclear and there are ways to incorporate big data into existing theories. But, in order to understand big data, a discourse about their relation towards organizational theories is required and essential for moving big data towards a more basic theoretical comprehension.

One important aspect of HRM is to close the research to practice gap (Huselid 2011) and, consequently, allow us to envisage it as an applied research stream. Thereby, big data force HRM to redefine itself extensively. At first it will be essential to research the importance of big data in HRM. As stated in chapter 2.4 there are two polarizing views and both are not beneficial for HRM. Consequently, aspects like the augmentation view as well as overcoming the HR-IT barrier need research to find ways to achieve these aspects. Research on electronic HRM will become even more important, however, not only on the operative perspective in digitizing the work of HRM. HRM will focus more and more on strategic work and less on operative work. This is a chance to demonstrate the strategic capabilities of HRM in dealing with the implementation of new technologies and especially with big data. Big data are rooted in the social level and, therefore, HRM will need to create solutions for this new situation. Another aspect that is linked with HRM is intercultural management. Instruments and methods may work in a distinct cultural environment, but will not work in different countries. Furthermore, culture is not solely linked to nations or regions, but also with other cultural influences. In the context of big data, some will have a strong impact on the big data within an organization, such as hacker culture (Scholz & Reichstein 2015), gamer culture (Stein & Scholz 2016), and IT culture (Scholz 2012). Big data will not make intercultural management obsolete; quite the opposite, intercultural management will influence big data within organizations considerably.

Furthermore, HRM will look for novel solutions for their implementation of big data. Big data transform the organization and the working world of every employee, however, big data do not present sufficient ideas for the implementation and usage of big data. Gamification and augmentation, for example, are less often utilized in the context of HRM. Gamification can be categorized as an emergent trend within HRM, although most applications are just off the shelf and detached from organizations. Gamification could potentially be far more than just some operative thing. I introduced a form of radical gamification, and with trained employees and a working HR daemon and HR centaur, people will tinker with big data and their working environment. Everything becomes more dynamic and more customized. Augmentation and applications for augmented reality will influence the working world in the future even more. Technologies like Magic Leap will mix reality and the digital world even further. It may be possible, in the near future, to have a complete fusion of the digital overlay with the real world. That increases the possibilities of big data drastically and may fundamentally change the way we work. Technology and especially big data open all possibilities, but HRM will be the research stream that develops fitting concepts for this new working world.

5.5 Implications for Teaching

Big data are part of our social world, but at the moment big data are often only taught in specialized data science degrees. Due to the comprehensive impact of big data and the proposed potential gain for everybody, having only some experts may not be sufficient. Above all, we are already facing a shortage of people who can deal with big data at a rudimentary level (Ahalt & Kelly 2013). The big data literacy proposed earlier will be part of the general studies of every curriculum (Lane 2016). Its extent can vary from the basics to in-depth knowledge.

This big data literacy will vary slightly from the big data literacy that is provided within an organization. The focus is on creating a long-term foundation in dealing with big data. The first objective is to overcome the big data phobia (Telgheder & Brower-Rabinowitsch 2016) many people are suffering from, especially as the current perception within society tends to be slightly negative due to stories about governmental over-surveillance (Jagadish et al. 2014). That is one element which needs to be discussed in the curriculum, so as to have an informed perspective on the current big data landscape rather than relying on media interpretation. If big data are part of general studies, people who fear numbers or hate mathematics will resist such teachings. It is essential to take the irrational fears of those students and reveal to them that big data can be fun, but at the very least, something they need to deal with, as big data will be part of their future jobs.

The next element is about the development of computational thinking. There are already several people (e.g. Wing 2006, Lee et al. 2011) who demand more efforts in teaching computational thinking in all academic fields. Wing (2006: 33) highlights the claim that "computational thinking is a fundamental skill for everyone, not just for computer scientists". The goal is, thus, to teach students the logic behind programming (Shein 2014), but this does not necessarily require them to learn programming (Conti 2006). "Thinking like a computer scientist means more than being able to program a computer. It requires thinking at multiple levels of abstraction" (Wing 2006: 34). Consequently, computational thinking will be a critical element in times of big data and digitization, but it grows from the fundamental logic behind computers and big data as well as the tools to deal with big data from an intellectual perspective. Without computational thinking, big data will stay in the black box and even look like something magical. Such computational thinking will be beneficial for potential radical gamification and for the fusion of the digital world and real world through augmentation.

Statistical thinking is, as the name suggests, about the ability to understand and conduct statistical analyses (Wild 1994). It can be defined as follows: "Statistical thinking involves an understanding of why and how statistical investigations are conducted and the "big ideas" that underlie statistical investigations. These ideas include the omnipresent nature of variation and when and how to use appropriate methods of data analysis such as numerical summaries and visual displays of data" (Ben-Zvi & Garfield 2005: 7). In times of big data, such skills are more needed than ever before and there is a need to build a solid foundation of statistical competence for any student in the future (Wild & Pfannkuch 1999). Big data will be driven by statistics and, although there is some complexity involved in statistics (Greer 2000), without any extensive knowledge of statistics people will be steered by a data-driven algorithm.

The final component in big data literacy is the skeptical thinking suggested by Sagan (1996), though this can be compared to critical thinking (Ennis 1962). Big data are complex, subjective, and influential. It may be difficult to understand big data and their impact meaning that it will be especially relevant to teach the idea of challenging big data analyses and questioning decisions given by big data. People will need a skill-set to ask the right questions (Davenport 2013) and request alternatives (Barton & Court 2012). Skeptical thinking is not seen as the constant questioning of big data, but the ability to think of new methods and to think outside the box. This creative thinking is realistic, however, and not a simulation of pipe dreams. Big data can simulate everything probable, but not everything is possible. The algorithms in particular will depend on skeptical thinking. They depend on the human ability to be reasonable and to challenge algorithms (Mainzer 2015).

Although many researchers (e.g. Frey & Osborne 2013, Brynjolfsson & McAfee 2014) expect that big data or new technologies in general will make several jobs obsolete, it becomes evident that some skills will be more important and others will be less relevant. In a study by the World Economic Forum (Gray 2016), the following skills are seen as essential in the future: complex problem solving, critical thinking, creativity, people management, coordinating with others, emotional intelligence, judgment and decision making, service orientation, negotiation, and cognitive flexibility. All these skills are demanded by a data-augmented homeodynamic organization and are required by all employees to a certain degree, but mostly for the HR department. The curriculum in HRM will find a way to train their students in big data literacy and teach them the competences for these skills.

5.6 Outlook

Big data are here to stay and we will deal with them. Big data will have an undeniable impact on society, organization, and the individual, however, it seems that big data will not become more objective over time, but rather blurrier with time (Bendler et al. 2014). Big data are influenced by one data-constructivism and external social-constructivism and those effects will become more influential over time. With every iteration the distortion will become more fossilized (Scholz 2015a). There are two developments in computing that may intensify the impreciseness of big data further. On the one hand, there is approximate computing (Mittal 2016), in which the computer will tolerate some loss in correctness in order to become quicker or more energy efficient (Han & Orshansky 2013). Big data will not be exact. On the other hand, there is quantum computing, the deviation from the binary principle and the introduction of the uncertainty principle (Heisenberg 1927). Although these computers will be faster and more secure (Preskill 1998), there will be an even more complex black box, putting more pressure on the question of accountability in this new data world (Nissenbaum 1996).

This thesis is a theoretical work, but hopefully it will spark further discussions. Big data will transform our organizations and the working world. Research is required to contribute a theoretical understanding of these changes. Big data are complex and organizations are facing an increasing complexity in their environment. Big data may be a young phenomenon, but its social impact is already hefty and challenges current research at every level.

Mainzer (2015) coined the German term *Technikgestaltung* (shaping) and focused on the idea that big data will be shaped and designed in a certain way. This shaping will be influenced by big data, but people are already working on the shaping of big data within organizations. Big data help to build smart factories (James 2012), smart offices (Le Gal et al. 2001), improve the working environment and support employees. As in the other industrial revolutions, people will find a way to harness the power, acquire new skills and adapt their work to the new situation. It will be interesting to see how big data augment the work of all employees and are potentially able to give employees more space to unfold their talents. Big data, designed in the right way, will give employees the freedom to data-source operational tasks and focus on more creative and innovative tasks.

A homeodynamic organization will face an increase in tension through big data and transformational power, however, big data can help to deal with those tensions. A data-augmented homeodynamic organization can be compared to the following statement: "Step in the same river twice? We don't even have that river anymore. We paved it over back when the rain stopped falling" (Wilson 2015). Organizations will become more dynamic and face more drastic changes from the outside than ever before. The world is perhaps not more dynamic than in the past, but, big data enable every organization to share *all* available information in nearly real-time. Such interaction between data will constantly create new challenges for the homeodynamic organization. Through this interaction, there will be constant new challenges, and those challenges will remain shrouded within big data. The advice of Morin will be more relevant than ever before:

> "Don't forget that reality is changing, don't forget that something new can (and will) spring up" (2008: 56).

Big data will continue to be a strong influence on organizations and on society, however, there is a real chance that they will be beneficial at the end, at least within organizations. Currently, technological progress envisions a future where the human factor will become the competitive advantage that organizations will keep in order to differentiate from other organizations. The success of Apple is an example. It does not derive from technological progress any longer, many other smartphones are far superior. Rather, it is how Apple makes their product special and how they attract their customers. The people working for Apple make reliable products, their customers trust Apple, and the marketing department is outstanding. Why is it that people are still making the difference in today's world? The way big data are used depends on people, but big data will enable people to do a better job and focus on their creative and innovational potential. The source of competitive advantage in times of big data will be the people within organizations. People will matter more as they are augmented by big data and can then focus on their core talents. At least for big data within organizations, I think the words of Douglas Adams are quite fitting: "Don't Panic!" (Adams, 2009: 6).

01000100 01101111 01101110 00100111 01110100 00100000 01010000 01100001 01101110 01101001 01100011 00100001

References

Aakster, Laurens L., & Keur, Ron (2012). Big data: Too big to ignore. What organizations can learn from the American presidential elections. Compact, 2: 1–8.

Adami, Christoph (1995). Self-organized criticality in living systems. Physics Letter A, 203(1): 29–32.

Adams, Douglas (2009). The hitchhiker's guide to the galaxy. London: Macmillan.

Adams, Pauline A., & Adams, Joe K. (1960). Confidence in the recognition and reproduction of words difficult to spell. American Journal of Psychology, 73(4): 544–552.

Adee, Sarah (2008). The data: 37 Years of Moore's Law. IEEE Spectrum, 45(5): 56–56.

Adler, Mortimer J. (1986). A guidebook to learning: For a lifelong pursuit of wisdom. London: Macmillan.

Agrawal, Divyakant, El Abbadi, Amr, Arora, Vaibhav, Budak, Ceren, Georgiou, Theodore, Mahmoud, Hatem A., Nawab, Faisal, Sahin, Cetin, & Wang, Shiyuan (2015). Mind your Ps and Vs: A perspective on the challenges of big data management and privacy concerns. International Conference on Big Data and Smart Computing (BigComp), Jeju, South Korea, 1–6.

Ahalt, Stan, & Kelly, Kip (2013). The big data talent gap. UNC Kenan-Flagler Business School White Paper, 1–15.

Aiden, Erez, & Michel, Jean-Baptiste (2013). Uncharted. Big data as a lens on human culture. New York: Riverhead.

Alavi, Maryam, & Leidner, Dorothy E. (2001). Knowledge management and knowledge management systems: Conceptual foundations and research issues. MIS Quarterly, 25(1): 107–136.

Aldrich, Howard (1979). Organizations and environments. Englewood Cliffs, NJ: Prentice Hall.

Aldrich, Howard, McKelvey, Bill, & Ulrich, Dave (1984). Design strategy from the population perspective. Journal of Management, 10(1): 67–86.

Allan, George (2003). A critique of using grounded theory as a research method. Electronic Journal of Business Research Methods, 2(1): 1–10.

Allison, Graham T. (1969). Conceptual models and the Cuban missile crisis. American Political Science Review, 63(3): 689–718.

Alvesson, Mats, & Kärreman, Dan (2011). Decolonializing discourse: Critical reflections on organizational discourse analysis. Human Relations, 64(9): 1121–1146.

Amaral, Luis A. N., & Uzzi, Brian. (2007). Complex systems-a new paradigm for the integrative study of management, physical, and technological systems. Management Science, 53(7): 1033–1035.

Amendola, Aniello (2002). Recent paradigms for risked informed decision making. Safety Science, 40(1):17–30.

Amin, Ash (1994). Post-Fordism: A reader. Oxford: Blackwell.

Amoore, Louise, & Piotukh, Volha (2015). Life beyond big data: Governing with little analytics. Economy and Society, 44(3): 341–366.

Andersen, P. Bøgh (1994). The semiotics of autopoiesis. A catastrophe-theoretic approach. Cybernetics & Human Knowing, 2(4): 17–38.

Andersen, Ross (2015). The most mysterious star in our galaxy, http://www.theatlantic.com/science/archive/2015/10/the-most-interesting-star-in-our-galaxy/410023/, last accessed 25 April 2016.

Anderson, Chris (2008). The end of theory: The data deluge makes the scientific method obsolete, http://archive.wired.com/science/discoveries/magazine/16-07/pb_theory, last accessed 26 April 2016.

Anderson, Chris, & Sally, David (2013). The numbers game: Why everything you know about football is wrong. London: Penguin.

Anderson, Janna Q., & Rainie, Lee (2012). Big data: Experts say new forms of information analysis will help people be more nimble and adaptive, but worry over humans' capacity to understand and use these new tools well. Washington, D.C.: Pew Research Center.

Anderson, Philip (1999). Perspective: Complexity theory and organization science. Organization Science, 10(3): 216–232.

Anderson, Philip., Meyer, Alan, Eisenhardt, Kathleen, Carley, Kathleen, & Pettigrew, Andrew (1999). Introduction to the special issue: Applications of complexity theory to organization science. Organization Science, 10(3): 233–236.

Andrejevic, Mark (2014). Big data, big questions: The big data divide. International Journal of Communication, 8: 1673–1689.

Andriani, Pierpaolo, & Cohen, Jack (2013). From exaptation to radical niche construction in biological and technological complex systems. Complexity, 18(5): 7–14.

Andriani, Pierpaolo, & McKelvey, Bill (2009). Perspective-from Gaussian to Paretian thinking: causes and implications of power laws in organizations. Organization Science, 20(6): 1053–1071.

Andriopoulos, Constantine, & Lewis, Marianne W. (2009). Exploitation-exploration tensions and organizational ambidexterity: Managing paradoxes of innovation. Organization Science, 20(4): 696–717.

Angrave, D., Charlwood, A., Kirkpatrick, I., Lawrence, M., & Stuart, M. (2016). HR and analytics: why HR is set to fail the big data challenge. Human Resource Management Journal, 26(1): 1–11.

Ansolabehere, Stephen, & Hersh, Eitan (2012). Validation: What big data reveal about survey misreporting and the real electorate. Political Analysis, 20(4): 4, 437–459.

Aradau, Claudia, & Blanke, Tobias (2015). 'The (big) data-security assemblage: Knowledge and critique'. Big Data & Society, 2(2): 1–12.

Arel, Itamar, Rose, Derek C.& Karnowski, Thomas P. (2010). Deep machine learning – A new frontier in artificial intelligence research. IEEE Computational Intelligence Magazine, 5(4): 13–18.

Argyris, Chris, & Schön, Donald A. (1996). Organizational learning: Theory, method and practice. Reading, MA: Addison-Wesley.

Arjoon, Surendra (2005). Corporate governance: An ethical perspective. Journal of Business Ethics, 61(4): 343–352.

Arkin, Ronald C. (1990). The impact of cybernetics on the design of a mobile robot system: A case study. IEEE Transactions on Systems, Man and Cybernetics, 20(6): 1245–1257.

Armenakis, Achilles A., Harris, Stanley G., & Mossholder, Kevin W. (1993). Creating readiness for organizational change. Human Relations, 46(6): 681–703.

Armstrong, Michael (2014). Armstrong's handbook of human resource management. London: Kogan Page.

Armstrong, J. Scott (2012). Predicting job performance: The Moneyball factor. Foresight: The International Journal of Applied Forecasting, 25(Spring): 31–34.

Arnott, David (2006). Cognitive biases and decision support systems development: A design science approach. Information Systems Journal, 16(1): 55–78.

Ashby, W. Ross (1952). Design for a brain. New York: Wiley.

Ashby, W. Ross (1956). An introduction to cybernetics. London: Methuen.

Asimov, Isaac (1951). Foundation. New York: Gnome Press.

Asimov, Isaac (2000). Foundation. Foundation and empire. Second foundation. New York: Everyman's Library.

Atzmueller, Martin, Ernst, Andreas, Krebs, Friedrich, Scholz, Christoph, & Stumme, Gerd (2016). Formation and temporal evolution of social groups during coffee breaks. 5th International Workshop on Mining Ubiquitous and Social Environments. Revised Selected Papers, 90–108.

Azuma, Ronald T. (1997). A survey of augmented reality. Presence: Teleoperators and Virtual Environments, 6(4): 355–385.

Baehr, Peter (2001). The "iron cage" and the "shell as hard as steel": Parsons, Weber, and the Stahlhartes Gehäuse metaphor in the Protestant ethic and the spirit of capitalism. History and Theory, 40(2): 153–169.

Bager, Jo (2006). Der Datenkrake. c't – magazin, 23: 168–171.

Baggaley, Jon (2010). The luddite revolt continues. Distance Education, 31(3): 337–343.

Bagley, Philip R. (1968). Extension of programming language concepts. Philadelphia: University City Science Center.

Bainbridge, William S. (2006). Cyberimmortality: Science, religion, and the battle to save our souls. The Futurist, 40(2): 25–29.

Bak, Per (1996). How nature works. The science of self-organized criticality. New York: Copernicus.

Bak, Per, Tang, Chao, & Wiesenfeld, Kurt (1988). Self-organized criticality. An explanation of 1/f noise. Physical Review Letters, 59(4): 381–384.

Bak, Per, Tang, Chao, & Wiesenfeld, Kurt (1988). Self-organized criticality. Physical Review A, 38(1): 364–374.

Baker, Chris (2008). Meet Leland Chee, the Star Wars franchise continuity cop, http://www.wired.com/2008/08/ff-starwarscanon, last accessed 25 April 2016.

Baker, Ted, & Nelson, Reed. E. (2005). Creating something from nothing: Resource construction through entrepreneurial bricolage. Administrative Science Quarterly, 50(3): 329–366.

Bakir, Vian (2015). Veillant panoptic assemblage: Mutual watching and resistance to mass surveillance after Snowden. Media and Communication, 3(3): 12–25.

Bar-Hillel, Maya (1980). The base-rate fallacy in probability judgments. Acta Psychologica, 44(3): 211–233.

Barabási, Albert-László (2003) Linked: How everything is connected to everything else and what it means for business, science, and everyday life. London: PLUME.

Barabási, Albert-László (2012). The network takeover. Nature Physics 8(1): 14–16.

Barabási, Albert-Lázsló (2013). Scientists must spearhead ethical use of big data, http:// www.politico.com/story/2013/09/scientists-must-spearhead-ethical-use-of-big-data-97578.html, last accessed 25 April 2016.

Barbaro, Michael, & Zeller, Tom (2006). A face is exposed for AOL searcher Nr. 4417749. http://www.nytimes.com/2006/08/09/technology/09aol.html?_r=1&, last accessed 16 February 2016.

Barlow, Mike (2013). Real-time big data analytics: Emerging architecture. Cambridge: O'Reilly.

Barrat, James (2013). Our final invention: Artificial intelligence and the end of the human era. New York: Thomas Dunne Books.

Barry, Bruce, & Stewart, Greg L. (1997). Composition, process, and performance in self-managed groups: The role of personality. Journal of Applied Psychology, 82(1): 62–78.

Barry, Christine A., Stevenson, Fiona A., Britten, Nicky, Barber, Nick, & Bradley, Colin P. (2001). Giving voice to the lifeworld. More humane, more effective medical care? A qualitative study of doctor-patient communication in general practice. Social Science and Medicine, 53(4): 487–505.

Barton, Dominic, & Court, David (2012). Making advanced analytics work for you. Harvard Business Review, 90(10): 78–83.

Bassler, Richard, A., & Joslin, Edward O. (1976). Managing data processing. Alexandria, VA: College Readings.

Bates, Alan P., & Cloyd, Jerry S. (1956). Toward the development of operations for defining group norms and member roles. Sociometry, 19(1): 26–39.

Baudrillard, Jean (1994). Simulucra and simulation. Ann Arbor: Michigan University Press.

Baudrillard, Jean, & Gane, Mike (1993). Baudrillard live: Selected interviews. London: Routledge.

Bauman, Zygmunt (2000). Liquid modernity. Cambridge: Polity Press.

Baumer, Benjamin, & Zimbalist, Andrew (2014). The sabermetric revolution: Assessing the growth of analytics in baseball. Philadelphia: University of Pennsylvania Press.

Baumgärtel, Tilmann (2015). Rebell aus Prinzip. Wired, 1(2): 30–31.

Beck, Charlie, & McCue, Colleen (2009). Predictive policing: What can we learn from Wal-Mart and Amazon about fighting crime in a recession? Police Chief, 76(11): 18–24.

Becker, Brian E., & Huselid, Mark A. (2006). Strategic human resources management: Where do we go from here? Journal of Management, 32(6): 898–925.

Beckers, John J., & Schmidt, Henk G. (2001). The structure of computer anxiety: A six-factor model. Computers in Human Behavior, 17(1): 35–49.

Beer, David (2009). Power through the algorithm? Participatory web cultures and the technological unconscious. New Media & Society, 11(6): 985–1002.

Beer, Michael, Spector, Bert, Lawrence, Paul, Mills, D. Quinn, & Walton, Richard E. (1984). Managing human assets. New York: Free Press.

Beinhocker, Eric D. (1997). Strategy at the edge of chaos. McKinsey Quarterly, 33(1): 24–40.

Bell, Genevieve (2015). The secret life of big data. In Tom Boellstorff, & Bill Maurer (Eds.). Data, now bigger and better! (pp. 7–26). Chicago: Prickly Paradigm Press.

Bell, Gordon, Hey, Tony, & Szalay, Alex (2009). Beyond the data deluge. Science, 323(5919): 1297–1298.

Bellah, Robert N. (1959). Durkheim and history. American Sociological Review, 24(4): 447–461.

Bellman, Richard E. (1957). Dynamic programming. Princeton, NJ: Princeton University Press.

Bellow, Francis (1959). The 1960s: A forecast of technology. Fortune 59: 74–78.

Ben-Zvi, Davi, & Garfield, Joan (2005). Statistical literacy, reasoning, and thinking: Goals, definitions, and challenges. In: Dani Ben-Zvi, & Joan Garfield (Eds.). The challenge of developing statistical literacy, reasoning and thinking (pp. 3–16). New York: Kluwer.

Bendler, Johannes, Wagner, Sebastian, Brandt, Tobias, & Neumann, Dirk (2014). Informa-tionsunschärfe in Big Data. Wirtschaftsinformatik, 56(5): 303–313.

Bentham, Jeremy (1843). The works of Jeremy Bentham. Volume four. Edinburgh: William Tait.

Berger, Peter L., & Luckmann, Thomas (1966). The social construction of reality: A treatise in the sociology of knowledge. New York: Anchor Books.

Bernerth, Jeremy B., Taylor, Shannon G., Walker, H. Jack, & Whitman, Daniel S. (2012). An empirical investigation of dispositional antecedents and performance-related outcomes of credit scores. Journal of Applied Psychology, 97(2): 469–478.

Bernstein, Ethan (2014). The transparency trap. Harvard Business Review, 92(10): 58–66.

Bersin, Josh (2012). Big data in HR: Building a competitive talent analytics function – The four stages of maturity. Bersin White Paper, 1–84.

Bessen, James (2015). Scarce skills, not scarce jobs. http://www.theatlantic.com/business/archive/2015/04/scarce-skills-not-scarce-jobs/390789, last accessed 25 April 2016.

Beyer, Mark A., & Laney, Doug (2012). The importance of "big data": A definition. Stamford, CT: Gartner.

Biddle, Bruce. J. (2013). Role theory: Expectations, identities, and behaviors. New York: Academic Press.

Biehn, Neil (2013). The missing V's in big data: Viability and value. http://www.wired.com/insights/2013/05/the-missing-vs-in-big-data-viability-and-value/, last accessed 03 February 2016.

Biermann, Christoph (2007). Im Geheimlabor des Fußballs. http://www.spiegel.de/spiegel/print/d-51074754.html, last accessed 23 April 2016.

Biermann, Christoph (2015). Moneyball im Niemandsland. 11 Freunde, 15(6): 90–96.

Bijker, Wiebe E., Hughes, Thomas P., & Pinch, Trevor Pinch (Eds.) (1987). The social construction of technological systems. New directions in the sociology and history of technology. Cambridge: MIT Press.

Bijker, Wiebe E., Hughes, Thomas P., & Pinch, Trevor (1999). General introduction. In Wiebe E. Bijker, Thomas P. Hughes, & Trevor Pinch (Eds.). The social construction of technological systems. New directions in the sociology and history of technology (pp. 1–6). Cambridge: MIT Press.

Bikard, Michaël (2012). Simultaneous discoveries as a research tool: Method and promise. SSRN Working Paper: 1–33.

Bimber, Bruce (1994). Three faces of technological determinism. In Merritt R. Smith, & Leo Marx (Eds.). Does technology drive history? The dilemma of technological determinism (pp. 79–100). Cambridge: MIT Press.

Bimber, Oliver, & Raskar, Ramesh (2005). Spatial augmented reality: Merging real and virtual worlds. Boca Raton, FL: CRC Press.

Bladt, Jeff, & Filbin, Bob (2013). Know the difference between your data and your metrics. http://blogs.hbr.org/2013/03/know-the-difference-between-yo, last accessed 25 April 2016.

Blauner, Robert (1964). Alienation and freedom: The factory worker and his industry. Chicago: University of Chicago Press.

Bless, Herbert, Fiedler, Klaus, & Strack, Fritz (2004). Social cognition: How individuals construct social reality. Hove and New York: Psychology Press.

Boczkowski, Pablo J. (2004). The mutual shaping of technology and society in videotext newspapers: Beyond the diffusion and social shaping perspectives. Information Society, 20(4): 255–267.

Boellstorff, Tom (2015). Making big data, in theory. In Tom Boellstorff, & Bill Maurer (Eds.). Data, now bigger and better! (pp. 87–108). Chicago: Prickly Paradigm Press.

Bogost, Ian (2014). Why gamification is bullshit. In Steffen P. Walz, & Sebastian Deterding (Eds.). The gameful world: Approaches, issues, applications (pp. 65–79). Cambridge: MIT Press.

Boisot, Max, & McKelvey, Bill (2010). Integrating modernist and postmodernist perspectives on organizations: A complexity science bridge. Academy of Management Review, 35(3): 415–433.

Bolin, Göran, & Schwarz, Jonas A. (2015). Heuristics of the algorithm: Big Data, user interpretation and institutional translation. Big Data & Society, 2(2): 1–11.

Bollier, David (2010). The promise and peril of big data. Washington, DC: The Aspen Institute.

Bonabeau, Eric (2009). Decisions 2.0: The power of collective intelligence. MIT Sloan Management Review, 50(2): 45–52.

Booch, Grady (2014). The human and ethical aspects of big data. IEEE Software, 31(1):20–22.

Boolos, George, & Richard, Jeffrey (1974). Computability and logic. Cambridge: Cambridge University Press.

Borge-Holthoefer, Javier, Perra, Nicola, Gonçalves, Bruno, González-Bailón, Sandra, Arenas, Alex, Moreno, Yamir, & Vespignani, Alessandro (2016). The dynamics of information-driven coordination phenomena: A transfer entropy analysis. Science Advances, 2(4): 1–8.

Borgmann, Christine L. (2007). Scholarship in the digital age. Cambridge: MIT Press.

Borne, Kirk (2014). Top 10 big data challenges – A serious look at 10 big data V's. https://www.mapr.com/blog/top-10-big-data-challenges---serious-look-10-big-data-v's, last accessed 25 April 2016.

Bosco, Frank A., Aguinis, Herman, Field, James G., Pierce, Charles A., & Dalton, Dan R. (2015). Harking's threat to organizational research: Evidence from primary and meta-analytic sources. Personnel Psychology, published online before print.

Bostrom, Nick (2006). How long before superintelligence. Linguistic and Philosophical Investigations, 5(1): 11–30.

Bouchikhi, Hamid, & Kimberly, John R. (2003). Escaping the identity trap. MIT Sloan Management Review, 44(3): 20–26.

Boulding, Kenneth E. (1956). The image. Ann Arbor: University of Michigan Press.

Bourdieu, Pierre (1977). Outline of a theory of practice. Cambridge: Cambridge University Press.

Bourdieu, Pierre (1986). Distinction: A social critique of the judgement of taste. London: Routledge.

Bovis, Beth, Pressman, Adam, Gagne, Dan, & Sisco, Braxton (2012). Quantitative talent management: A Moneyball perspective. Human Resource Executive Online, 1–4.

Bowker, Geoffrey C. (2005). Memory practices in science. Cambridge: MIT Press.

Bowker, Geoffrey C. (2014). The theory/data thing. International Journal of Communication, 8: 1795–1799.

Boxer, Philip, & Kenny, Vincent (1990). The economy of discourses: a third order cybernetics? Human Systems Management, 9(4): 205–224.

Boyajian, Tabetha S., LaCourse, Daryll. M., Rappaport, Saul A., Fabrycky, Daniel, Fischer, Debra A., Gandolfi, Davide, Kennedy, Gareth M., Korhonen, Heidi, Liu, Molin C., Moor, Atilla, Olah, Katalin, Vida, Krisztián, Wyatt, Mark C., Best, William M. J., Brewer, John M., Ciesla, Fred J., Csak, Balázs, Deeg, Hans J., Dupuy, Trent J., Handler, Gerald, Heng, Kevin, Howell, Steve B.,Ishikawa, Sasha T., J. Kovacs, Kozakis, Thea, Kriskovics, Levente, Lehtinen, Jyri, Lintott, Chris, Lynn, Stuart, Nespral, David, Nikbakhsh, Shabnam, Schawinski, Kevin, Schmitt, Joseph R., Smith, Arfon M., Szabo, Gy M., Szabo, Róbert, Viuho, J., Wang, Jin-Qing, Weiksnar, Alex, Bosch, Mike, Connors, J. L., Goodman, Samuel J., Green, Gerald R., Hoekstra, Abe J., Jebson, Tony, Jek, Kian J., Omohundro, Mark R., Schwengeler, Hans M., Szewczyk, A. (2016). Planet hunters IX. KIC 8462852-Where's the Flux? Monthly Notices of the Royal Astronomy Society, 457(4): 3988–4004.

Boyd, Danah, & Crawford, Kate (2012). Critical questions for big data. Provocations for a cultural, technological, and scholarly phenomenon. Information, Communication & Society 15(5): 662–679.

Bradford, David L., & Burke, W. Warner (Eds.) (2005). Reinventing organization development: New approaches to change in organizations. San Francisco: Pfeiffer.

Branch, Melville C. (1997). Simulation, planning, and society. Westport: Praeger.

Brenner, Peter J. (2013). Herrschaft durch Zahlen. universitas, 48: 34–53.

Brillouin, Leon (1953). The negentropy principle of information. Journal of Applied Physics, 24(9): 1152–1163.

Brin, David (2012). Fiction: Ray Bradbury, an appreciation. Nature, 486(7404): 471.

Briscoe, Gerard, & Mulligan, Catherine (2014). Digital innovation: The hackathon phenomenon. London: Creativeworks.

Brosnan, Mark J. 2002. Technophobia: The psychological impact of information technology. London: Routledge.

Brown, Victoria R., & Vaughn, E. Daley (2011). The writing on the (Facebook) wall: The use of social networking sites in hiring decisions. Journal of Business and Psychology, 26(2): 219–225.

Bryant, Adam (2011). Google's quest to build a better boss. http://www.nytimes.com/2011/03/13/business/13hire.html, last accessed 26 April 2016.

Bryant, Levi R. (2011). The democracy of objects. Ann Arbor: Open Humantities Press.

Brynjolfsson, Erik, Hitt, Lorin M., & Kim, Heekyung Hellen (2011). Strength in numbers: How Does data-driven decisionmaking Affect firm performance? SSRN Working Paper: 1–33.

Brynjolfsson, Erik, & McAfee, Andrew (2011). Race against the machine. Lexington, MA: Digital Frontier Press.

Brynjolfsson, Erik, & McAfee, Andrew (2014). The second machine age. New York: W. W. Norton.

Bryson, Steve, Kenwright, David, Cox, Michael, Ellsworth, David, & Haimes, Robert (1999). Visually exploring gigabyte data sets in real time. Communications of the ACM, 42(8): 82–90.

Buchanan, Mark (2015). Physics in finance: Trading at the speed of light. Nature, 518(7538): 161–163.

Buchhorn, Eva (2014). App als Chef – wie Software Mitarbeiter durchleuchtet, http://www.manager-magazin.de/magazin/artikel/personalmanagement-software-durchleuchtet-mitarbeiter-a-1022736.html, last accessed 25 April 2016.

Buckingham, Marcus (2015). Most HR data is bad data, https://hbr.org/2015/02/most-hr-data-is-bad-data, last accessed 12 September 2015.

Buhl, Hans U., Röglinger, Maximilian, & Moser, Florian (2013). Big data. A fashionable topic with(out) sustainable relevance for research and practice? Wirtschaftsinformatik, 55(2): 63–68.

Bunge, Mario (1963). A general black box theory. Philosophy of Science, 30(4): 346–358.

Burke, Alafair S. (2007). Neutralizing cognitive bias: An invitation to prosecutors. N.Y.U. Journal of Law & Liberty, 2: 512–530.

Burrell, Jenna (2016). How the machine 'thinks': Understanding opacity in machine learning algorithms. Big Data & Society, 3(1): 1–12.

Busck, Ole, Knudsen, Herman, & Lind, Jens (2010). The transformation of employee participation: Consequences for the work environment. Economic and Industrial Democracy, 31(3): 285–305.

Bylander, Tom, Allemang, Dean, Tanner, Michael C., & Josephson, John R. (1991). The computational complexity of abduction. Artificial Intelligence, 49(1): 25–60.

Byrnes, Nanette (2016). Disruptive Zeiten bei Künstlicher Intelligenz, http://www.heise.de/tr/artikel/Disruptive-Zeiten-bei-Kuenstlicher-Intelligenz-3164251.html, last accessed 25 April 2016.

Calás, Marta B., & Smircich, Linda (1999). Past postmodernism? Reflections and tentative directions. Academy of Management Review, 24(4): 649–672.

Caldwell, Raymond (2001). Champions, adapters, consultants and synergists: the new change agents in HRM. Human Resource Management Journal, 11(3): 39–52.

Campbell, Donald T. (1979). Assessing the impact of planned social change. Evaluation and Program Planning, 2(1): 67–90.

Cannon, Walter B. (1926). Physiological regulation of normal states: some tentative postulates concerning biological homeostatics. In Paris Academy of Medicine (Ed.). Jubilee volume to Charles Richet (pp. 91–93). Paris: Editions Medicales.

Cannon, Walter B. (1929). Organization for physiological homeostasis. Physiological Reviews, 9(3): 399–431.

Cappelli, Peter (2015). Why we love to hate HR ... and what HR can do about it. Harvard Business Review, 93(7/8): 54–61.

Carr, Nicholas (2014). The limits of social engineering, https://www.technologyreview.com/s/526561/the-limits-of-social-engineering, last accessed 08 March 2016.

Carroll, Lewis (1991). Through the looking-glass and what Alice found there. https://www.gutenberg.org/files/12/12-h/12-h.htm, last accessed 23 April 2016.

Castells, Manuel (2010). The information age. Economy, society and culture. Volume I: The rise of the network society. Chichester: Wiley-Blackwell.

Cassidy, Mike (2014). Centaur chess shows power of teaming human and machine. http://www.huffingtonpost.com/mike-cassidy/centaur-chess-shows-power_b_6383606.html, last accessed 25 April 2016.

Chakrabarti, Soumen (2009). Data mining: Know it all. Burlington, MA: Morgan Kaufmann.

Chan, Anita (2015). Big data interfaces and the problem of inclusion. Media, Culture & Society, 37(7): 1078–1083.

Charan, Ram, Barton, Dominic, & Carey, Dennis (2015). People before strategy: A new role for the CHRO. Harvard Business Review, 93(7/8): 62–71.

Charnley Gail, Elliott, E. Donald (2002). Risk versus precaution: Environmental law and public health protection. Environmental Law Reporter, 32(3):10363–10366.

Chen, Hsinchun, Chiang, Roger H. L., & Storey, Veda C. (2012). Business intelligence and analytics: From big data to big impact. MIS Quarterly, 36(4): 1165–1188.

Cheng, Bing, & Titterington, D. Michael (1994). Neural networks: A review from a statistical perspective. Statistical Science, 9(1): 2–30.

Chesbrough, Henry W. (2006). Open innovation: The new imperative for creating and profiting from technology. Cambridge: Harvard Business Press.

Child, John, Ihrig, Martin, & Merali, Yasmin (2014). Organization as information – A space odyssey. Organization Studies, 35(6): 801–824.

Chomsky, Noam (1965). Aspects of the theory of syntax. Cambridge: MIT Press.

Chomsky, Noam (2002). Media control: The spectacular achievements of propaganda, New York: Seven Stories Press.

Chomsky, Noam (2013). Noam Chomsky and Bart Gellman at engaging data, http://www.hyperorg.com/blogger/2013/11/15/liveblog-noam-chomsky-at-engaging-data, last accessed 06 January 2016.

Christensen-Szalanski, Jay J., & Willham, Cynthia. F. (1991). The hindsight bias: A meta-analysis. Organizational Behavior and Human Decision Processes, 48(1): 147–168.

Christozov, Dimitar, & Toleva-Stoimenova, Stefka. (2015). Big data literacy: A new dimension of digital. In John Girard, Deanna Klein, & Kristi Berg (Eds.). Strategic data-based wisdom in the big data era (pp. 156–171). Hershey, PA: Information Science Reference.

Cilliers, Paul (1998). Complexity and postmodernism: Understanding complex systems. London: Routledge.

CIPD (2013). Talent analytics and big data – The challenge for HR. London: CIPD.

Clark, Andy (2015). You are what you eat. In John Brockman (Ed.). What to think about machines that think (pp. 156–159). New York: Harper Perennial.

Clarke, Arthur C. (1977). Profiles of the future. New York: Popular Library.

Clarke, Juanne N. (1981). A multiple paradigm approach to the sociology of medicine, health and illness. Sociology of Health & Illness, 3(1): 89–103.

Clemons, Eric K. (2013). Online profiling and invasion of privacy: The myth of anonymization. http://www.huffingtonpost.com/eric-k-clemons/internet-targeted-ads_b_2712586.html, last accessed 16 February 2016.

Clemons, Eric. K., Wilson, James, & Jin, Fujie (2014). Investigations into consumers' preferences concerning privacy: An initial step towards the development of modern and consistent privacy protections around the globe. 47th Hawaii International Conference on System Science (HICSS), Waikoala, 4083–4092.

Coase, Ronald. H. (1937). The nature of the firm. economica, 4(16): 386–405.

Cohen, Michael D., March, James G., & Olsen, Johan P. (1972). A garbage can model of organizational choice. Administrative Science Quarterly, 17(1): 1–25.

Collingridge David (1996). Resilience, flexibility, and diversity in managing the risks of technologies. In Christopher Hood, & David K. C. Jones (Eds.). Accidents and design: Contemporary debates in risk management (pp. 40–45). Oxford: Routledge.

Conrad, Klaus (1958). Die beginnende Schizophrenie. Versuch einer Gestaltanalyse des Wahns. Stuttgart: Georg Thieme.

Conti, Gregory (2006): Introduction. Communications of the ACM, 49(6): 33–36.

Cook, Samantha, Conrad, Corrie, Fowlkes, Ashley L., & Mohebbi, Matthew H. (2011). Assessing Google flu trends performance in the United States during the 2009 influenza virus A (H1N1) pandemic. PloS one, 6(8): 1–8.

Cooper, Robert, & Burrell, Gibson (1988). Modernism, postmodernism and organizational analysis: An introduction. Organization Studies, 9(1): 91–112.

Cormen, Thomas H., Leiserson, Charles E., Rivest, Ronald L., & Stein, Clifford (2009). Introduction to algorithms. Cambridge: MIT Press.

Cornerstone OnDemand (2013). Big data in HR. Santa Monica, CA: Cornerstone OnDemand.

Cox, Michael, & Ellsworth, David (1997). Managing big data for scientific visualization. ACM Siggraph, 97: 1–17.

Craik, Kenneth J. W. (1943). The nature of explanation. Cambridge: Cambridge University Press.

Crail, Mark (2015). The problem with the sorcery of big data in HR.http://www.managers.org.uk/insights/news/2015/august/the-problem-with-the-sorcery-of-big-data-in-hr, last accessed 23 April 2016.

Crawford, Kate (2013). The hidden biases in big data. https://hbr.org/2013/04/the-hidden-biases-in-big-data, last accessed 11 February 2016.

Crawford, Kate, Miltner, Kate, & Gray, Mary L. (2014). Critiquing big data: Politics, ethics, epistemology. International Journal of Communication, 8: 1663–1672.

Csikszentmihalyi, Mihaly (2008). Flow: The psychology of happiness. New York: Harper Collins.

Cukier, Kenneth (2013). Dehumanising human resources. http://www.economist.com/blogs/schumpeter/2013/04/big-data-and-hiring, last accessed 26 April 2016.

Cumbley, Richard, & Church, Peter (2013). Is "big data" creepy? Computer Law & Security Review, 29(5): 601–609.

D'Ignazio, Catherine, & Bhargava, Rahul (2015). Approaches to big data literacy. Bloomberg Data for Good Exchange Conference, New York, 1–6.

Dahl, Richard (2010). Green washing: Do you know what you're buying? Environmental Health Perspectives, 118(6): 246–252.

Dalton, Craig, & Thatcher, Jim (2014). What does a critical data studies look like, and why do we care? Seven points for a critical approach to 'big data'. http://societyandspace.com/material/commentaries/craig-dalton-and-jim-thatcher-what-does-a-critical-data-studies-look-like-and-why-do-we-care-seven-points-for-a-critical-approach-to-big-data, last accessed 09 February 2016.

Darwin, Charles (1859). On the origin of species. London: John Murray.

Dasgupta, Sayamindu, & Resnick, Mitchel (2014). Engaging novices in programming, experimenting, and learning with data. ACM Inroads, 5(4): 72–75.

Dass, Parshotam, & Parker, Barbara (1999). Strategies for managing human resource diversity: From resistance to learning. Academy of Management Executive, 13(2): 68–80.

Data Processing Management Association (DPMA) (1970). Data management. Minneapolis: Charles Babbage Institute.

Davenport, Thomas H. (2006). Competing on analytics. Harvard Business Review, 84(1): 98–107.

Davenport, Thomas H. (2013). Keep up with the quants. Harvard Business Review, 91(7/8): 120–123.

Davenport, Thomas H. (2014). Big data at work: Dispelling the myths, uncovering the opportunities. Boston: Harvard Business Review Press.

Davenport, Thomas H., & Kirby, Julia (2015). Beyond automation. Harvard Business Review, 93(6): 58–65.

Davenport, Thomas H., & Patil, D. J. (2012). Data scientist. The sexiest job of the 21st century. Harvard Business Review, 90(10): 70–76.

David, Paul A. (1985). Clio and the Economics of QWERTY. The American economic review, 75(2): 332–337.

Davis, Kord (2012). Ethics of big data. Sebastopol, CA: O'Reilly.

de Biase, Luca (2015). Narratives and our civilization. In John Brockman (Ed.). What to think about machines that think (pp. 231–234). New York: Harper Perennial.

de Goes, John (2013). Big data is dead. What's next? http://venturebeat.com/2013/02/22/big-data-is-dead-whats-next, last accessed 03 February 2016.

de Mauro, Andrea, Greco, Marco, & Grimaldi, Michele (2015). What is big data? A consensual definition and a review of key research topics. AIP Conference Proceedings, Madrid, 97–104.

de Montjoye, Yves-Alexandre, Radaelli, Laura, & Singh, Vivek K. (2015). Unique in the shopping mall: On the reidentifiability of credit card metadata. Science, 347(6221): 536–539.

de Vries, Gerard, Verhoeven, Imrat, & Boeckhout, Martin (2011). Taming uncertainty: The WRR approach to risk governance. Journal of Risk Research, 14(4): 485–499.

Degele, Nina (2002). Einführung in die Techniksoziologie. München: Fink.

Deng, Li, & Yu, Dong (2014). Deep learning: Methods and applications. Foundations and Trends in Signal Processing, 7(3–4): 197–387.

Derose, Chris (2013). How Google uses data to build a better worker. http://www.theatlantic.com/business/archive/2013/10/how-google-uses-data-to-build-a-better-worker/280347, last accessed 25 April 2016.

Desai, Preyas S. (2001). Quality segmentation in spatial markets: When does cannibalization affect product line design? Marketing Science, 20(3): 265–283.

Deterding, Sebastian, Sicart, Miguel, Nacke, Lennart, O'Hara, Kenton, & Dixon, Dan (2011). Gamification. Using game-design elements in non-gaming contexts. Proceedings of the Conference on Human-Computer Interaction, Vancouver, 2425–2428.

Deterding, Sebastian (2014). Eudaimonic design, or: Six invitations to rethink gamification. In Mathias Fuchs, Sonia Fizek, Paolo Ruffino, & Niklas Schrape (Eds.). Rethinking gamification (pp. 305–331). Lüneburg: meson press.

Deutsch, David (2002). The structure of the multiverse. Proceedings of the Royal Society of London A: Mathematical, Physical and Engineering Sciences, 458(2028): 2911–2923.

Dewhurst, Martin, & Willmott, Paul (2014). Manager and machine: The new leadership equation. McKinsey Quarterly, http://www.mckinsey.com/global-themes/leadership/manager-and-machine, last accessed 23 April 2016.

deWinter, Jennifer, Kocurek, Carly A., & Nichols, Randall (2014). Taylorism 2.0: Gamification, scientific management and the capitalist appropriation of play. Journal of Gaming & Virtual Worlds, 6(2): 109–127.

Diebold, Francis X. (2000). Big data dynamic factor models for macroeconomic measurement and forecasting, 8th World Congress of the Econometric Society, 115–122.

Diebold, Francis X. (2012). On the origin(s) and development of the term "big data". PIER Working Paper, 12-037: 1–6.

Diesner, Jana (2015). Small decisions with big impact on data analytics. Big Data & Society, 2(2): 1–6.

Dietsch, Jeanne (2010). People meeting robots in the workplace. IEEE Robotics & Automation Magazine, 17(2): 15–16.

Dijcks, Jean-Pierre (2012). Oracle. Big data for enterprise. Oracle Working Paper, 1–14.

Diller, Ann (1997). In praise of objective-subjectivity: Teaching the pursuit of precision. Studies in Philosophy and Education, 16(1–2): 73–87.

Dilley, Roy (1999). Introduction: The problem of context. In Roy Dilley (Ed.). The problem of context (pp. 1–46). New York: Berhahn.

DiMaggio, Paul J., & Powell, Walter W. (1983). The iron cage revisited: Institutional isomorphism and collective rationality in organizational fields. American Sociological Review, 48(2): 147–160.

Dinov, Ivo D., Petrosyan, Petros, Liu, Zhizhong, Eggert, Paul, Zamanyan, Alen, Torri, Federica, Macciardi, Fabio, Hobel, Sam, Woo Moon, Seok, Hee Sung, Young, Jiang, Zhiguo, Labus, Jennifer, Kurth, Florian, Ashe-McNalley, Cody, Mayer, Emeran, Vespa, Paul M., Van Horn, John D., & Toga, Arthur W. (2014). The perfect neuroimaging-genetics-computation storm: Collision of petabytes of data, millions of hardware devices and thousands of software tools. Brain Imaging and Behavior, 8(2): 311–322.

Dix, Alexander (2016). Datenschutz im Zeitalter von Big Data. Wie steht es um den Schutz der Privatsphäre? Stadtforschund und Statistik, 26(1): 59–64.

Dobelli, Rolf (2015). Self-aware AI? Not in 1,000 years! In John Brockman (Ed.). What to think about machines that think (pp. 98–101). New York: Harper Perennial.

Doctorow, Cory (2015). Reputation economy dystopia: China's new "citizen scores" will rate every person in the country, https://boingboing.net/2015/10/06/reputation-economy-dystopia-c.html, last accessed 25 April 2016.

Dodge, Martin, & Kitchin, Rob (2005). Codes of life: Identification codes and the machine-readable world. Environment and Planning D: Society and Space, 23(6): 851–881.

Dorminey, Bruce (2014). Harpoon malfunction may have saved ESA's Philae comet lander.http://www.forbes.com/sites/brucedorminey/2014/11/30/harpoon-malfunction-may-have-saved-esas-philae-comet-lander/#5d4fb0433237, last accessed 25 April 2016.

Dörner, Dietrich (1989): Die Logik des Mißlingens. Strategisches Denken in komplexen Situationen. Reinbek bei Hamburg: Rowohlt.

Dougherty, Dale (2012). The maker movement. innovations, 7(3): 11–14.

Dourish, Paul (2004) What we talk about when we talk about context. Personal and Ubiquitous Computing, 8(1): 19–30.

Dourish, Paul, & Bell, Genevieve (2013). 'Resistance is futile': Reading science fiction alongside ubiquitous computing. Personal and Ubiquitous Computing, 18(4): 769–778.

Drengson, Alan R. (1984). The sacred and the limits of the technological fix. Zygon, 19(3): 259–275.

Dreyfus, Hubert (1965). Alchemy and AI. Santa Monica, CA: RAND Corporation.

Drucker, Peter F. (1967). The manager and the moron. McKinsey Quarterly. http://www.mckinsey.com/business-functions/organization/our-insights/the-manager-and-the-moron, last accessed 23 April 2016.

Drumm, Hans J., & Scholz, Christian (1988). Personalplanung. Planungsmethoden und Methodenakzeptanz. Bern – Stuttgart: Haupt.

Duhigg, Charles (2012). How companies learn your secrets. http://www.nytimes.com/2012/02/19/magazine/shopping-habits.html, last accessed 25 April 2016.

Dumbill, Edd (2013). Making sense of big data. Big Data, 1(1): 1–2.

Duncan, Robert B. (1976). The ambidextrous organization: Designing dual structures for innovation. In Ralph H. Kollman, Louis R. Pondy, & Dennis Sleven (Eds.). The management of organization (pp. 167–188). New York: North Holland.

Durkheim, Emile (1897/1951). Suicide. Glencoe, IL: Free Press.

Dutcher, Jennifer (2014). What is big data? https://datascience.berkeley.edu/what-is-big-data, last accessed 03 February 2016.

Duymedjian, Raffi, & Rüling, Charles-Clemens (2010). Towards a foundation of bricolage in organization and management theory. Organization Studies, 31(2): 133–151.

Dwork, Cynthia., & Mulligan, Deirdre K. (2013). It's not privacy, and it's not fair. Stanford Law Review Online, 66: 35–40.

Dyche, Jill (2012). Big data "Eurekas!" don't just happen. https://hbr.org/2012/11/eureka-doesnt-just-happen, last accessed 09 February 2016.

Eagle, Nathan, & Pentland, A. Sandy (2006). Reality mining: Sensing complex social systems. Personal and Ubiquitous Computing, 10: 255–268.

Economist (2013). Robot recruiters, http://www.economist.com/node/21575820, last accessed 26 April 2016.

Economist (2014). Planet of the phones. The smartphone is ubiquitous, addictive and transformative. http://www.economist.com/news/leaders/21645180-smartphone-ubiquitous-addictive-and-transformative-planet-phones, last accessed 23 April 2016.

Ehrhardt, Michelle (2016). What Tinder and Halo have in common, http://www.theatlantic.com/entertainment/archive/2016/01/how-tinder-matchmaking-is-like-warcraft/424350, last accessed 25 April 2016.

Eisenhardt, Kathleen M. (1989). Building theories from case study research. Academy of Management Review, 14(4): 532–550.

Eisenhardt, Kathleen. M., & Martin, Jeffrey A. (2000). Dynamic capabilities: What are they? Strategic Management Journal, 21(10–11): 1105–1121.

Ekbia, Hamid, Mattioli, Michael, Kouper, Inna, Arave, G., Ghazinejad, Ali, Bowman, Timothy, Suri, Venkata R., Tsou, A., Weingart, Scott, & Sugimoto, Cassidy R.

(2015). Big data, bigger dilemmas: A critical review. Journal of the Association for Information Science and Technology, 66(8): 1523–1545.

Elkan, Charles (2001). Magical thinking in data mining: Lessons from CoIL challenge 2000. Proceedings of the Seventh ACM SIGKDD International Conference on Knowledge Discovery and Data Mining, San Francisco, 426–431.

Ellenberg, Jordan (2014). What's even creepier than Target guessing that you're pregnant? http://www.slate.com/blogs/how_not_to_be_wrong/2014/06/09/big_data_what_s_even_creepier_than_target_guessing_that_you_re_pregnant.html, last accessed 25 April 2016.

Ellul, Jacques (1964). The technological society. New York: Vintage Books.

Emerson, John W., & Kane, Michael J. (2012). Don't drown in the data, Significance, 9(4): 38–39.

Emery, Fred E., & Trist, Eric L. (1965). The causal texture of organizational environments. Human Relations, 18(1): 21–32.

Ennis, Robert H. (1962). A concept of critical thinking. Harvard Educational Review, 32(1): 81–111.

Enserink, Martin, & Chin, Gilbert (2015). The end of privacy. Science, 347(6221): 490–491.

Eoyang, Glenda H. (2007). Human system dynamics: Complexity-based approach to a complex evaluation. In Bob Williams, & Iraj Iman (Eds.). Systems concepts in evaluation: An expert anthology (pp. 75–88). Point Reyes: Edge Press.

Eoyang, Glenda H. (2011) Complexity and the dynamics of organizational change. In Peter Allen, Steve Maguire, & Bill McKelvey (Eds.) The SAGE handbook of complexity and management (pp. 317–333). Thousand Oaks: Sage.

Eppler, Martin J., & Mengis, Jeanne (2004). The concept of information overload: A review of literature from organization science, accounting, marketing, MIS, and related disciplines. The Information Society, 20(5): 325–344.

Epstein, Susan L. (2015). Wanted: Collaborative intelligence. Artificial Intelligence, 221: 36–45.

eQuest (2013). Big data: HR's golden opportunity arrives. San Ramon, CA: eQuest.

Esposti, Sara Degli (2014). When big data meets dataveillance: The hidden side of analytics. Surveillance & Society, 12(2): 209–225.

Evans, Paul A., & Doz, Yves (1992). Dualities: A paradigm for human resource and organizational development in complex multinationals. In: Vladimir Pucik, Noel M. Tichy, & Carole K. Barnett (Eds.). Globalizing management: Creating and leading the competitive organization (pp. 85–106). New York: Wiley.

Eve Online (2003). Reyknjavík: CCP Games.

Evolv (2013). Q3 2013 Workforce performance report. Evolv White Paper: 1–20.

Falconer, James (2002). Emergence happens! Misguided paradigms regarding organizational change and the role of complexity and patterns in the change landscape. Emergence, 4(1–2): 117–130.

Falconer, Kenneth (1997). Techniques in fractal geometry. Hoboken, NJ: Wiley.

Falkvinge, Rick (2015). In China, your credit score is now affected by your political opinions – and your friends' political opinions. https://www.privateinternetaccess.com/blog/2015/ 10/in-china-your-credit-score-is-now-affected-by-your-political-opinions-and-your-friends-political-opinions, last accessed 25 April 2016.

Fanning, Kurt, & Centers, David P. (2013). Intelligent business process management: Hype or reality? Journal of Corporate Accounting & Finance, 24(5): 9–14.

Farjoun, Moshe (2010). Beyond dualism: Stability and change as a duality. Academy of Management Review, 35(2): 202–225.

Fauver, Larry, & Fuerst, Michael E. (2006). Does good corporate governance include employee representation? Evidence from German corporate boards. Journal of Financial Economics, 82(3): 673–710.

Fayyad, Usama M., Piatetsky-Shapiro, Gregory, & Smyth, Padhraic (1996). Knowledge discovery and data mining: Towards a unifying framework. KDD, 96: 82–88.

Federal Trade Commission (FTC) (2012). Report to Congress under Section 319 of the Fair and Accurate Credit Transactions Act of 2003, 1–370.

Federal Trade Commission (FTC) (2015). Report to Congress under Section 319 of the Fair and Accurate Credit Transactions Act of 2003, 1–83.

Feffer, Mark (2015). Processing people. HR Magazine, 60(8): 40–45.

Feist, Richard, Beauvais, Chantal, & Shukla, Rajesh (2010). Introduction. In Richard Feist, Chantal Beauvais, & Rajesh Shukla (Eds.). Technology and the changing face of humanity (pp. 1–21). Ottawa: Ottawa University Press.

Feldman, Lauren, Maibach, Edward, W., Roser-Renouf, Connie, & Leiserowitz, Anthony (2012). Climate on cable: The nature and impact of global warming coverage on Fox News, CNN, and MSNBC. International Journal of Press/Politics, 17(1): 3–31.

Feldman, Maryann P., & Francis, Johanna L. (2004). Homegrown solutions: Fostering cluster formation. Economic Development Quarterly, 18(2): 127–137.

Fellows, Michael R., Fomin, Fedor V., Lokshtanov, Daniel, Rosamond, Frances, Saurabh, Saket, & Villanger, Yngve (2012). Local search: Is brute-force avoidable? Journal of Computer and System Sciences, 78(3): 707–719.

Finley, Klint (2013). Mathematicians predict the future with data from the past, http://www.wired.com/2013/04/cliodynamics-peter-turchin, last accessed 25 April 2016.

Fischhoff, Baruch, & Beyth, Ruth (1975). 'I knew it would happen': Remembered probabilities of once-future things. Organizational Behaviour and Human Performance, 13(1): 1–16.

Fisher, Danyel, DeLine, Rob, Czerwinski, Mary, & Drucker, Steven (2012). Interactions with big data analytics. interactions, 19(3): 50–59.

Floridi, Luciano. (2012). Big data and their epistemological challenge. Philosophy & Techno-logy, 25(4): 435–437.

Fombrun, Charles J., Tichy, Noel, M., & Devanna, Mary A. (1984). Strategic human resource management. Hoboken, NJ: Wiley.

Ford, Jeffrey D., Ford, Laurie W., & D'Amelio, Angelo (2008). Resistance to change: The rest of the story. Academy of Management Review, 33(2): 362–377.

Ford, Martin (2015). Rise of the robots. Technology and the threat of a jobless future. New York: Basic Books.

Forrester, Jay W. (1971). Counterintuitive behavior of social systems. Theory and Decision, 2(2): 109–140.

Forrester, Jay W. (1994). System dynamics, systems thinking, and soft OR. System Dynamics Review, 10(2–3): 245–256.

Forsyth, John, & Boucher, Leah (2015). Why big data is not enough. Research World, 2015(50): 26–27.

Foucault, Michel (1977). Discipline and punish: The birth of the prison. New York: Pantheon Books.

Fowler, Alastair (1979). Genre and the literary canon. New Literary History, 11(1): 97–119.

Frankel, Felice, & Reid, Rosalind (2008). Big data: Distilling meaning from data. Nature, 455(7209): 30.

Franklin, Ursula M. (1999). The real world of technology. Concord, ON: House of Anansi.

Freeman, John, Carroll, Glenn R., & Hannan, Michael T. (1983). The liability of newness: Age dependence in organizational death rates. American Sociological Review, 48(5): 692–710.

Freeman, Richard (2007). Epistemological bricolage: How practitioners make sense of learning. Administration & Society, 39(4): 476–496.

Freiberger, Paul, & Swaine, Michael (1999). Fire in the valley: Making of the Personal Computer. New York: McGraw-Hill.

Frey, Carl B., & Osborne, Michael A. (2013). The future of employment: How susceptible are jobs to computerisation? OMS Working Paper, 1–72.

Frické, Martin (2015). Big data and its epistemology. Journal of the Association for Information Science and Technology, 66(4): 651–661.

Friendly, Michael (2008). A brief history of data visualization. In Chun-Houh Chen, Wolfgang Karl Härdle, & Antony Unwin (Eds.). Handbook of data visualization (pp. 15–56). Berlin: Springer.

Frischmann, Brett M. (2014). Human-focused Turing tests: A framework for judging nudging and techno-social engineering of human beings. Cardozo Legal Studies Research Paper, 441: 1–58.

Froese, Tom (2010). Life after Ashby: Ultrastability and the autopoietic foundations of biological autonomy. Cybernetics & Human Knowing, 17(4): 7–49.

Froomkin, A Michael (2015). From anonymity to identification. Journal of Self-Regulation and Regulation, 1: 121–138.

Gadamer, Hans-Georg (1992). Truth and method. New York: Crossroad.

Galagan, Pat (2014). HR gets analytical. T+D, 68: 22–25.

Gallagher, Kelly (2006). Rethinking the Fair Credit Reporting Act: When requesting credit reports for employment purposes goes too far. Iowa Law Review, 91: 1593–1621.

Gallagher, Sean (2013). What the NSA can do with 'big data'. http://www.arstechn ica.com /information-technology/2013/06/what-the-nsa-can-do-with-big-data, last accessed 26 April 2016.

Gandomi, Amir, & Haider, Murtaza (2015). Beyond the hype: Big data concepts, methods, and analytics. International Journal of Information Management, 35(2): 137–144.

George, Gerry, Haas, Martine R., & Pentland, A. Sandy (2014). Big data and management. Academy of Management Journal, 57(2): 321–326.

Gerhart, Barry (2005). Human resources and business performance: Findings, unanswered questions, and an alternative approach. Management Revue, 16(2): 174–185.

Giard, François, & Guitton, Matthieu J. (2016). Spiritus ex machina: Augmented reality, cyberghosts and externalised consciousness. Computers in Human Behavior, 55(B): 614–615.

Gibson, Cristina B., & Birkinshaw, Julian (2004). The antecedents, consequences, and mediating role of organizational ambidexterity. Academy of Management Journal, 47(2): 209–226.

Giddens, Anthony (1979). Central problems in social theory: Action, structure, and contradiction in social analysis. Berkeley: University of California Press.

Giddens, Anthony (1984) The constitution of society: Outline of the theory of structuration. Polity Press, Cambridge.

Giddings, Seth (2006). Walkthrough: Videogames and technocultural form. PhD thesis, University of West of England.

Gigerenzer, Gerd (2015). Robodocters. In John Brockman (Ed.). What to think about machines that think (pp. 317–320). New York: Harper Perennial.

gild (2013). The big data recruiting playbook. San Francisco: gild.

Gillespie, Tarleton (2012). Can an algorithm be wrong? Limn, 1(2), http://limn.it/can-an-algorithm-be-wrong, last accessed 27 April 2016.

Ginsberg, Jeremy, Mohebbi, Matthew H., Patel, Rajan S., Brammer, Lynette, Smolinski, Mark S., & Brilliant, Larry (2009) Detecting influenza epidemics using search engine query data. Nature 457(7232): 1012–1014.

Gitelman, Lisa (Ed.) 2013. "Raw data" is an oxymoron. Cambridge: MIT Press.

Gitelman, Lisa, & Jackson, Virginia (2013). Introduction. In Lisa Gitelman (Ed.) 2013. "Raw data" is an oxymoron (pp. 1–14). Cambridge: MIT Press.

Glaser, Barney G., & Strauss, Anselm L. (1967) The discovery of grounded theory: strategies for qualitative research. Piscataway, NJ: Aldine.

Glassmeier, Karl-Heinz, Boehnhardt, Hermann, Koschny, Detlef, Kührt, Ekkehard, & Richter, Ingo (2007) The Rosetta mission: Flying towards the origin of the solar system. Space Science Reviews 128(1–4):1–21.

Goffman, Erving (1959). The presentation of self in everyday life. New York: Anchor Books.

Goldberg, David E., & Holland, John H. (1988). Genetic algorithms and machine learning. Machine learning, 3(2): 95–99.

Gomez, Peter, & Probst, Gilbert (1980). Centralization versus decentralization in business organizations: cybernetic rules for effective management. Cybernetics and Systems, 11(4): 381–400.

González, Marta C., Hidalgo, Cesar A., & Barabási, Albert-László (2008). Understanding individual human mobility patterns. Nature, 453(7196): 779–782.

Gore, Al (2013). The future: Six drivers of global change. New York: Random House.

Government Office for Science (2016). Distributed ledger technology: beyond block chain, https://www.gov.uk/government/uploads/system/uploads/attachment_data/file/492972/gs-16-1-distributed-ledger-technology.pdf, last accessed 26 April 2016.

Graham, Mark, Zook, Matthew, & Boulton, Andrew (2013). Augmented reality in urban places: contested content and the duplicity of code. Transactions of the Institute of British Geographers, 38(3): 464–479.

Graham, Shawn (2013). It's time to rethink human resources, the key to employee morale. http://www.fastcompany.com/3002355/its-time-rethink-human-resources-key-employee-morale, last accessed 26 April 2016.

Gray, Alex (2016). The 10 skills you need to thrive in the fourth industrial revolution. https://www.weforum.org/agenda/2016/01/the-10-skills-you-need-to-thrive-in-the-fourth-industrial-revolution, last accessed 25 April 2016.

Green, Leila (2001). Technoculture. Crows Nest: Allen and Unwin.

Greene, Brian (2005). Making sense of string theory. https://www.ted.com/talks/brian_greene_ on_string_theory/transcript?language=en, last accessed 24 April 2016.

Greene, Brian (2016). Did we just discover aliens? Scientists aren't ruling it out. http://ideas.ted.com/did-we-just-discover-aliens, last accessed 25 April 2016.

Greene, William H. (2003). Econometric analysis. Upper Saddle River, NJ: Prentice Hall.

Greer, Brian (2000). Statistical thinking and learning. Mathematical Thinking and Learning, 2(1–2): 1–9.

Gregory, Robert W., & Muntermann, Jan (2011). Theorizing in design science research: Inductive versus deductive approaches. 32nd International Conference on Information Systems, Shanghai, 1–17.

Griffiths, Tom (2015). Brains and other thinking machines. In: John Brockman (Ed.). What to think about machines that think (pp. 141–144). New York: Harper Perennial.

Grimes, Sara M. (2006). Online multiplayer games: A virtual space for intellectual property debates. New Media & Society, 8(6): 969–990.

Grimson, William E. L. (1980). Computing shape using a theory of human stereo vision. PhD Thesis. Massachusetts Institute of Technology.

Gros, Claudius (2012). Pushing the complexity barrier: Diminishing returns in the sciences. Complex Systems, 21(3): 183–192.

Gross, David (1982). Time-space relations in Giddens' social theory. Theory, Culture & Society, 1(2): 83–88.

Grossmann, Bettina, Lames, Martin. & Stefani, Ray (2015). From talent to professional football-youthism in German football. International Journal of Sports Science and Coaching, 10(6): 1103–1114.

Guertin, Carolyn (2012). Digital prohibition: Piracy and authorship in new media art. New York: Bloomsbury.

Guszcza, James, Steier, David, Lucker, John, Gopalkrishnan, Vivekanand, & Lewis, Harvey (2013). Too big to ignore. Deloitte Review, 12: 36–53.

Guthrie, Cameron (2014). Empowering the hacker in us: a comparison of fab lab and hackerspace ecosystems. 5th Latin American and European Meeting on Organization Studies Colloquium (LAEMOS), 2–5.

Habermas, Jürgen (1970). Toward a rational society: Student protest, science and politics. Boston: Beacon Press.

Hagendijk, Rob, & Irwin, Alan (2006) Public deliberation and governance: Engaging with science and technology in contemporary Europe. Minerva 44(2):167–184.

Haggerty, Kevin D. & Ericson, Richard V. (2000). The surveillant assemblage. British Journal of Sociology, 51(4): 605–622.

Hamari, Juho, Koivisto, Joanna, & Sarsa, Harri (2014). Does gamification work – A literature review of empirical studies on gamification. 47th Hawaii International Conference on Systems Sciences (HICSS), Waikoala, 1–10.

Hamel, Gary (1998). Path breaking. Executive Excellence, 15(1): 3–4.

Hampton, Stephanie E., Strasser, Carly A., Tewksbury, Joshua J., Gram, Wendy K., Budden, Amber E., Batcheller, Archer L., Duke, Clifford S., & Porter, John H. (2013). Big data and the future of ecology. Frontiers in Ecology and the Environment, 11(3): 156–162.

Han, Jiawei, Kamber, Micheline, & Pei, Jian (2012). Data mining: Concepts and techniques. Waltham, MA: Morgan Kaufmann.

Han, Jie, & Orshansky, Michael (2013). Approximate computing: An emerging paradigm for energy-efficient design. 18th IEEE European Test Symposium (ETS), 1–6.

Hand, David J. (1998). Data mining: Statistics and more? The American Statistician, 52(2): 112–118.

Hand, David J. (2016). Editorial: 'Big data' and data sharing. Journal of the Royal Statistical Society: Series A (Statistics in Society), 179(3): 629–631.

Hannan, Michael T., & Freeman, John (1977). The population ecology of organizations. American Journal of Sociology, 82(5): 929–964.

Hannan, Michael T., & Freeman, John (1984) Structural inertia and organizational change. American Sociological Review, 49(2): 149–164.

Hanson, Norwood R. (1958). The logic of discovery. The Journal of Philosophy, 55(25): 1073–1089.

Harford, Tim (2014). Big data: Are we making a big mistake? Significance, 11(5):14–19.

Harris, Mark (2015) Documents confirm Apple is building self-driving car, http://www.theguardian.com/technology/2015/aug/14/apple-self-driving-car-project-titan-sooner-than-expected, last accessed 19 October 2015.

Harris, Jeanne (2012). Data is useless without the skills to analyze it. https://hbr.org/2012/09/data-is-useless-without-the-skills, last accessed 25 April 2016.

Hartley, Peter, & Chatterton, Peter (2001). Business communication. Rethinking your professional practice for the post-digital age. New York: Routledge.

Harvard Business Review (2013). The big data opportunity for HR and finance. Cambridge: Harvard Business Review.

Hawking, Stephen, Russell, Stuart, Tegmark, Max, & Wilczek, Frank (2014). Stephen Haw-king: 'Transcendence looks at the implications of artificial intelligence – but are we taking AI seriously enough?' http://www.independent.co.uk/news/science/stephen-hawking-transcendence-looks-at-the-implications-of-artificial-intelligence-but-are-we-taking-9313474.html, last accessed 08 February 2016.

Hawley, Amos (1968). Human ecology. In David L. Sills (Ed.). International encyclopedia of the social sciences (pp. 328–337). New York: Macmillan.

Hayden, Michael (2014). Former NSA boss: "We kill people based on metadata". https://www.youtube.com/watch?v=UdQiz0Vavmc, last accessed 16 February 2016.

Hearst, Marti (2015). eGaia, a distributed technical-social mental system. In John Brockman (Ed.). What to think about machines that think (pp. 280–281). New York: Harper Perennial.

Heidegger, Martin (1977). The question concerning technology and other essays. New York: Harper & Row.

Heidegger, Martin (2011). Gesamtausgabe. II. Abteilung: Vorlesungen (1919–1944). Frankfurt am Main: Vittorio Klostermann.

Heilbroner, Robert L. (1967). Do machines make history. Technology and Culture, 8(3): 335–345.

Heisenberg, Werner (1927). Über den anschaulichen Inhalt der quantentheoretischen Kinematik und Mechanik. Zeitschrift für Physik, 43 (3–4): 172–198.

Helbing, Dirk (2015). Thinking ahead – Essays on big data, digital revolution, and participatory market society. Heidelberg: Springer.

Helland, Pat (2011). If you have too much data, then 'good enough' is good enough. Communications of the ACM, 54(6): 40–47.

Hellekson, Karen, & Busse, Kristina (2006). Fan fiction and fan communities in the age of the internet. Jefferson, NC: McFarland.

Hempel, Carl G. (1965). Typological methods in the natural and social sciences. In Carl G. Hempel (Ed.). Aspects of scientific explanation and other essays in the philosophy of science (pp. 155–171). New York: Free Press.

Hendler, Jim (2013). Broad data: Exploring the emerging web of data. Big Data, 1(1):18–20.

Hershbach, Dennis R. (1995). Technology as knowledge: Implications for instruction. Journal of Technology Education, 7(Fall): 31–42.

Hey, Tony, Tansley, Stewart, & Tolle, Kristin (2009). Jim Gray on eScience: A transformed cientific method. In: Tony Hey, Stewart Tansley, & Kristin Tolle (Eds.). The fourth paradigm. Data-intensive discovery (pp. xvii-xxxi). Redmond, WA: Microsoft Research.

Hilbert, Martin (2016). Big data for development: A review of promises and challenges. Development Policy Review, 34(1): 135–174.

Hilbert, Martin, & López, Priscila (2011). The world's technological capacity to store, communicate, and compute information. Science, 332(6025): 703–705.

Hildebrandt, Mireille (2013). Slaves to big data. Or are we? IDP. Revista de Internet, Derecho y Politica, 16, published online before print.

Hill, Kashmir (2012). How Target figured out a teen girl was pregnant before her father did. http://www.forbes.com/sites/kashmirhill/2012/02/16/how-target-figured-out-a-teen-girl-was-pregnant-before-her-father-did/#53e48efb34c6, last accessed 25 April 2016.

Hocquet, Alexander (2016). Football manager: Mutual shaping between, game, sport, and community. Kinephanos, 6(Special Issue): 34–52.

Hoenkamp, Eduard (2012). Taming the terabytes: A human-centered approach to surviving the information deluge. In Judith B. Strother, Jan M. Ulijn, & Zohra Fazal (Eds.). Information overload: An international challenge for professional engineers and technical communicators (pp. 145–174). Hoboken, NJ: Wiley.

Hoeren, Thomas (Ed.) (2014). Big Data und Recht. München: C. H. Beck.

Hoerl, Roger, & Snee, Ron D. (2012). Statistical thinking: Improving business performance. Hoboken, NJ: Wiley.

Holland, John H. (1995). Hidden order: How adaptation builds complexity. New York: Basic Books.

Hollinger, Richard C., & Adams, Amanda (2008). 2008 National Retail Security survey – Final report. Gainesville: University of Florida.

Holzkämpfer, Hendrik (1995). Management von Singularitäten und Chaos. Außergewöhnliche Ereignisse und Strukturen in industriellen Unternehmen. Wiesbaden: Springer.

Hopgood, Adrian A. (2003). Artificial intelligence: Hype or reality? Computer, 36: 24–28.

Hota, Chittaranjan, Upadhyaya, Shambhu, & Al-Karaki, Jamal N. (2015). Advances in secure knowledge management in the big data era. Information Systems Frontiers, 17(5): 983–986.

Hotten, Russell (2015). Carmakers face challenge from Google and Apple, http://www.bbc.com/news/business-31720645, last accessed 19 October 2015.

Hua, Chai (2015). Mainland credit-rating network takes shape. http://www.chinadaily asia.com/business/2015-06/09/content_15274221.html, last accessed 25 April 2016.

Hughes, Thomas P. (1969). Technological momentum in history: Hydrogenation in Germany 1898–1933. Past & Present, 44: 106–132.

Hughes, Thomas P. (1994). Technological momentum. In: Merritt R. Smith, & Leo Marx (Eds.). Does technology drive history. The dilemma of technological determinism (pp. 101–114). Cambridge: MIT Press.

Hume, David (1739). A treatise of human nature. Oxford: Clarendon Press.

Huotari, Kai & Hamari, Juho (2012). Defining gamification – A service marketing perspective. Proceedings of the 16th International Academic Mindtrek Conference, Tampere, Finland, 1–22.

Hurst, David K. (1995). Crisis and renewal: Ethical anarchy in mature organizations. Business Quarterly, 60(2): 32–32.

Huselid, Mark A. (2011). Celebrating 50 Years: Looking back and looking forward: 50 years of Human Resource Management. Human Resource Management, 50(3): 309–312.

Huselid Mark A. (2015). Workforce analytics for strategy execution. In Dave Ulrich, William A. Schiermann, & Libby Sartain (Eds.). The rise of HR: Wisdom from 73 thought leaders (pp. 301–315). Alexandria, VA: HR Certification Institute.

Hutter, Marcus (2009). Feature reinforcement learning: Part I. Unstructured MDPs. Journal of Artificial General Intelligence, 1(1): 3–24.

Huxley, Aldous (1932). Brave new world. London: Chatto & Windus.

Huxley, Aldous (1958). The Mike Wallace interview. http://www.hrc.utexas.edu/multimedia/video/2008/wallace/huxley_aldous_t.html, last accessed 26 April 2016.

Ilinitch, Anne Y., D'Aveni, Richard A., & Lewin, Arie Y. (1996). New organizational forms and strategies for managing in hypercompetitive environments. Organization Science, 7(3): 211–220.

Intel (2012). Big data analytics. Intel's IT manager survey on how organizations are using big data. Intel White Paper: 1–26.

Ioannidis, John P.A. (2005). Why most published research findings are false. PLoS Medicine, 2(8): 696–701.

IRGC (2005). Risk governance: Towards an integrative approach. IRGC White Paper 1:1–156.

Jackson, Brandon (2014). Helping HR get a seat at the table. https://rework.withgoogle.com/blog/getting-a-seat-at-the-table/, last accessed 25 April 2016.

Jackson, D. A., Somers, K. M. (1991). The spectre of 'spurious' correlations. Oecologia, 86(1): 147–151.

Jacobs, Adam (2009). The pathologies of big data. Communications of the ACM, 52(8): 36–44.

Jagadish, H. V., Gehrke, Johannes, Labrinidis, Alexandros, Papakonstantinou, Yannis, Patel, Jignesh M., Ramakrishnan, Raghu, & Shahabi, Cyrus (2014). Big data and its technical challenges. Communications of the ACM, 57(7): 86–94.

James, Timothy (2012). Smart factories. Engineering & Technology, 7(6): 64–67.

Janssens, Maddy, & Steyaert, Chris (1999). The world in two and a third way out? The concept of duality in organization theory and practice. Scandinavian Journal of Management, 15(2): 121–139.

Jensen, Howard E. (1952). Editorial note. In Howard P. Becker (Ed.). Through values to social interpretation. Essays on social contexts and prospects (pp. vii-xi). Durham: Duke University Press.

Johannessen, Jon-Arild, & Hauan, Arnulf (1994). Communication – A systems theoretical point of view (third-order cybernetics). Systems Practice, 7(1): 63–73.

John, Nicholas A. (2014). File sharing and the history of computing: Or, why file sharing is called "file sharing". Critical Studies in Media Communication, 31(3): 198–211.

Johnson-Laird, Philip N. (2004). The history of mental models. In Ken Manktelow, & Man Cheung Chung (Eds.). Psychology of reasoning (pp. 179–212). Hove-New York: Psychology Press.

Johnson, Jeffrey A. (2015). How data does political things: The processes of encoding and decoding data are never neutral. http://blogs.lse.ac.uk/impactofsocial sciences/2015/10/07/how-data-does-political-things/, last accessed 26 April 2016.

Johnson, Steven, (2007). The ghost map: The story of London's most terrifying epidemic – and how it changed science, cities, and the modern world. New York: Riverhead Books.

Johns, Gary (2006). The essential impact of context on organizational behavior. Academy of Management Review, 31(2): 386–408.

Jones, Chuck (2015). Hitting the brakes on Apple's electric car. http://www.forbes.com/sites/chuckjones/2015/02/14/hitting-the-brakes-on-apples-electric-car, last accessed 19 October 2015.

Junqué de Fortuny, Enric, Martens, David, & Provost, Foster (2013). Predictive modeling with big data: is bigger really better? Big Data, 1(4): 215–226.

Kafka, Franz (1925). Der Process. Berlin: Die Schmiede.

Kahaner, Larry (1997). Competitive intelligence. How to gather analyze, and use information to move your business to the top. New York: Touchstone.

Kahn, Jordan (2015) Revealed: The experts Apple hired to build an electric car. http://9to5mac.com/2015/02/19/apple-electric-car-team, last accessed 19 October 2015.

Kahneman, Daniel, & Lovallo, Dan (1993). Timid choices and bold forecasts: A cognitive perspective on risk taking. Management Science, 39(1): 17–31.

Kahneman, Daniel, & Tversky, Amos (1972). Subjective probability: A judgment of representativeness. Cognitive Psychology, 3(3): 430–454.

Kaisler, Stephen, Armour, Frank, Espinosa, J. Alberto, & Money, William (2013). Big data: Issues and challenges moving forward. 46th Hawaii International Conference on System Sciences (HICSS), Wailea, 995–1004.

Kast, Fremont E., & Rosenzweig, James E. (1972). General systems theory: Applications for organization and management. Academy of Management Journal, 15(4): 447–465.

Kasting, James F., Whitmire, Daniel P., & Reynolds, Ray T. (1993). Habitable zones around main sequence stars. Icarus, 101(1): 108–128.

Kates, Robert W. (1969). Mirror or monitor for man. Antipode, 1(1): 47–53.

Katz, Daniel, & Kahn, Robert L. (1966). The social psychology of organizations. New York: Wiley.

Kauffman, Stuart A. (1993). The origins of order: Self organization and selection in evolution. Oxford: Oxford University Press.

Kauffman, Stuart A. (1995). At home in the universe. The search for the laws of self-organization and complexity. Oxford: Oxford University Press.

Kaufman, Bruce E. (2015). Evolution of strategic HRM as seen through two founding books: A 30th anniversary perspective on development of the field. Human Resource Management, 54(3): 389–407.

Kebede, Gashaw (2010). Knowledge management: An information science perspective. International Journal of Information Management, 30(5): 416–424.

Kelly, Kevin (2014). The future of AI? Helping human beings think smarter. http://www.wired.co.uk/magazine/archive/2014/12/features/brain-power, last accessed 10 August 2015.

Kenny, Vincent (2009). There's nothing like the real thing. Revisiting the need for a third-order cybernetics. Constructivist Foundations, 4(2): 100–111.

Kerr, Norbert L. (1998). HARKing: Hypothesizing after the results are known. Personality and Social Psychology Review, 2(3): 196–217.

Khoury, Muin J., & Ioannidis, John P. A. (2014). Big data meets public health: Human well-being could benefit from large-scale data if large-scale noise is minimized. Science, 346(6213): 1054–1055.

Kidwell, Jillinda, Vander Linde, Karen M., & Johnson, Sandra L. (2000). Applying corporate knowledge management practices in higher education. EDUCAUSE Quarterly 23(4): 28–33.

Kilduff, Martin (2006). Editor's comments: Publishing theory. Academy of Management Review, 31(2): 252–255.

Kiley, Ellen P. (Ed.) (2005). World of Warcraft. The roleplaying game. Stone Mountain, GA: Sword and Sorcery.

Kim, Gang-Hoon, Trimi, Silvana, & Chung, Ji-Hyong (2014). Big-data applications in the government sector. Communications of the ACM, 57(3): 78–85.

Kim, Jeffrey, Lund, Arnie, & Dombrowski, Caroline (2013). Telling the story in big data. interactions, 20(3): 48–51.

Kim, Oliver W. (2014). The Moneyball myth. http://www.thecrimson.com/article/2014/3/10/the-moneyball-myth, last accessed 25 April 2016.

Kirilenko, Andrei A., Kyle, Albert S., Samadi, Mehrdad, & Tuzun, Tugkan (2015). The flash crash: The impact of high frequency trading on an electronic market. SSRN Working Paper, 1–41.

Kitchin, Rob (2013). Big data and human geography Opportunities, challenges and risks. Dialogues in Human Geography, 3(3): 262–267.

Kitchin, Rob (2014a). The data revolution. Big data, open data, data infrastructures & their consequences. Los Angeles: SAGE.

Kitchin, Rob (2014b). Big data, new epistemologies and paradigm shifts. Big Data & Society, 1(1): 1–12.

Kitchin, Rob (2014c). The real-time city? Big data and smart urbanism. GeoJournal, 79(1): 1–14.

Kitchin, Rob, & Dodge, Martin (2011). Code/Space: Software and everyday life. Cambridge: MIT Press.

Kitchin, Rob, & Lauriault, Tracey P. (2015). Towards critical data studies: Charting and unpacking data assemblages and their work. The Programmable City Working Paper 2, 1–19.

Klein, Hans K., & Kleinman, Daniel L. (2002). The social construction of technology: Structural considerations. Science, Technology & Human Values, 27(1): 28–52.

Kleinberg, Jon & Mullainathan, Sendhil (2015). We built them, but we don't understand them. In John Brockman (Ed.). What to think about machines that think (pp. 62–65). New York: Harper Perennial.

Kluckhohn, Florence R., & Strodtbeck, Fred L. (1961). Variations in value orientations. Evanston, IL. and Elmsford, NY: Row, Peterson.

Knop, Cartsten (2014): Dem deutschen Mittelstand ist die Digitalisierung egal. Frankfurter Allgemeine Zeitung, 211 (11. September): 25.

Koch, Tom (2004). The map as intent: variations on the theme of John Snow. Cartographica: The International Journal for Geographic Information and Geovisualization, 39(4): 1–14.

Kosko, Bart (2015). Thinking machines = old algorithms on faster computers. In John Brockman (Ed.). What to think about machines that think (pp. 423–426). New York: Harper Perennial.

Kosslyn, Stephen M. (2015). Another kind of diversity. In John Brockman (Ed.). What to think about machines that think (pp. 228–230). New York: Harper Perennial.

Kowalski, Robert (2011). Computational logic and human thinking. How to be artificially intelligent. Cambridge: Cambridge University Press.

Kranzberg, Melvin (1986). Technology and history: "Kranzberg's Laws". Technology and Culture, 27(3): 544–560.

Kraska, Tim (2013). Finding the Needle in the Big Data Systems Haystack. IEEE Internet Computing, 17(1): 84–86.

Krech, David, & Crutchfield, Richard S. (1948). Theory and problems of social psychology. New York: McGraw-Hill.

Krugman, Paul (1996). The self-organizing economy. Oxford: Blackwell.

Kucklick, Christoph (2014). Die granulare Gesellschaft. Wie das Digitale unsere Wirklichkeit auflöst. Berlin: Ullstein.

Kugler, Logan (2016). What happens when big data blunders? Communications of the ACM, 59(6): 15–16.

Kühl, Stefan (Eds.). Schlüsselwerke der Organisationsforschung. Wiesbaden: Springer VS.

Kuhn, Kristine M. (2013). What we overlook: Background checks and their implications for discrimination. Industrial and Organizational Psychology, 6(4): 419–423.

Kuhn, Thomas S. (1962). The structure of scientific revolutions. Chicago: University of Chicago Press.

Kull, Christoph (2016). Karriere im Algorithmus. Business Punk, 5(3): 13.

Kuner, Christopher, Cate, Fred H., Millard, Christopher, & Svantesson, Dan Jerker B. (2012). The challenge of 'big data' for data protection. International Data Privacy Law, 2(2): 47–49.

Kunz, William M. (2006). Culture conglomerates: Consolidation in the motion picture and tele-vision industries. New York: Rowman & Littlefield.

Kuper, Simon (2008). Milan Lab's secret of youth, http://www.ft.com/cms/s/0/c56b9be6-e6ff-11dc-b5c3-0000779fd2ac.html, last accessed 25 April 2016.

Knuth, Donald (2011). The art of programming. ITNOW, 53(4): 18–19.

Kurzweil, Ray (2006). The singularity is near: When humans transcend biology. New York: Viking.

LaFrance, Adrienne (2015). Not even the people who write algorithms really know how they work. http://www.theatlantic.com/technology/archive/2015/09/not-even-the-people-who-write-algorithms-really-know-how-they-work/406099/, last accessed 23 April 2016.

Lagoze, Carl (2014). Big data, data integrity, and the fracturing of the control zone. Big Data & Society, 1(2): 1–11.

Lane, Julia (2016). Big data: The role of education and training. Journal of Policy Analysis and Management, 35(3): 722–724.

Laney, Douglas (2001). 3D data management: Controlling data volume, velocity and variety, Stamford, CT: META Group.

Lange, Lydia L. (2002). The impact factor as a phantom: Is there a self-fulfilling prophecy effect of impact? Journal of Documentation, 58(2): 175–184.

Laplace, Pierre-Simon (1951). A philosophical essay on probabilities. New York: Dover.

Larose, Daniel T. (2014). Discovering knowledge in data: an introduction to data mining. Hoboken, NJ: Wiley.

Lashinsky, Adam (2016). Exclusive Q&A: Apple CEO Tim Cook. http://fortune.com/tim-cook-apple-q-and-a, last accessed 25 April 2016.

Latour, Bruno (1987). Science in action: How to follow scientists and engineers through society. Milton Keynes: Open University Press.

Latour, Bruno (1988). The pasteurization of France. Cambridge: Harvard University Press.

Latour, Bruno (1991). Technology is society made durable. In John Law (Ed.). A sociology of monsters: Essays on power, technology and domination (pp. 103–132). London: Routledge.

Latour, Bruno (1992). Where are the missing masses? The sociology of a few mundane artefacts. In Wiebe Bijker & John Law (Eds.). Shaping technology / Building society: studies in sociotechnical change (pp 225–258). Cambridge: MIT Press.

Latour, Bruno (1999). On recalling ANT. Sociological Review, 47(S1): 15–25.

Latour, Bruno (2002). Gabriel Tarde and the end of the social. In Patrick Joyce (Ed.). The social question. New bearings in history and the social science (pp. 117–132). London: Routledge.

Latour, Bruno (2005). Reassembling the social. An introduction to actor-network-theory. Oxford: Oxford University Press.

Latour, Bruno, Jensen, Pablo, Venturini, Tommaso, Grauwin, Sébastian, & Boullier, Dominique (2012). 'The whole is always smaller than its parts' – A digital test of Gabriel Tardes' monads. British Journal of Sociology 63(4): 590–615.

Law, John (1992). Notes on the theory of the actor-network-theory: Ordering, strategy, and heterogeneity. Systems Practice, 5(4): 379–393.

Lawson, Clive (2004). Technology, technological determinism and the transformational model of technical activity. Proceedings of the 2004 Annual Conference of the International Association for Critical Realism, Cambridge, UK, 1–24.

Lay, Dwane (2012). Why the WSJ is dead wrong about Moneyball, http://leanhrblog.com/why-the-wsj-is-dead-wrong-about-moneyball, last accessed 26 April 2016.

Layton, Edwin T. (1974). Technology as knowledge. Technology and Culture, 15(1): 31–41.

Lazer, David, Kennedy, Ryan, King, Gary, & Vespignani, Alessandro (2014). The parable of Google Flu: Traps in big data analysis. Science, 343 (6176):1203–1205.

Lazer, David, Pentland, A. Sandy, Adamic, Lada, Aral, Sinan, Barabási, Albert-Laszlo, Brewer, Devon, Christakis, Nicholas, Contractor, Noshir, Fowler, James, Gutmann, Myron, Jebara, Tony, King, Gary, Macy, Michael, Roy, Deb, & Van Alstyne, Marshall (2009). Life in the network: The coming age of computational social science. Science, 323(5915): 721–723.

Lazonder, Ard W., Biemans, Harm J. A., & Wopereis, Iwan G. J. H. (2000). Differences between novice and experienced users in searching information on the World Wide Web. Journal of the American Society for Information Science, 51(6): 576–581.

Le Gal, Cristophe, Martin, Jérôme, Lux, Augustin, & Crowley, James L. (2001). Smart office: Design of an intelligent environment. IEEE Intelligent Systems, 16(4): 60–66.

Leahu, Lucian, Schwenk, Steve, & Sengers, Phoebe (2008). Subjective objectivity: Negotiating emotional meaning. Proceedings of the 7th ACM Conference on Designing Interactive Systems, Cape Town, South Africa, 425–434.

Lee, Tim (2015). A place in the history books. IEEE Microwave Magazine, 16(2): 8–9.

Lee, Irene, Martin, Fred, Denner, Jil, Coulter, Bob, Allan, Walter, Erickson, Jeri, Malyn-Smith, Joyce, & Werner, Linda (2011). Computational thinking for youth in practice. ACM Inroads, 2(1): 32–37.

Lee, Yang, Madnick, Stuart, Wang, Richard, Wang, Forea, & Zhang, Hongyun (2014). A cubic framework for the chief data officer: Succeeding in a world of big data. MIS Quarterly Executive, 13(1): 1–13.

Lem, Stanislav (2013). Summa technologiae. Minnesota: Minnesota University Press.

Leonelli, Sabina (2014). What difference does quantity make? On the epistemology of Big Data in biology. Big Data & Society, 1(1): 1–11.

Lepak, David P., & Snell, Scott. A. (1998). Virtual HR: Strategic human resource management in the 21st century. Human Resource Management Review, 8(3): 215–234.

Lepak, David P., & Snell, Scott. A. (1999). The human resource architecture: Toward a theory of human capital allocation and development. Academy of Management Review, 24(1): 31–48.

Lepping, Peter (2011). Anticipatory obedience. The Psychiatrist, 35(7): 275–275.

Lessig, Lawrence (2008). Remix. London: Bloomsbury.

Levenson, Alec (2014). The promise of big data for HR. People & Strategy, 36: 22–26.

Lévi-Strauss, Claude (1966). The savage mind. Chicago: University of Chicago Press.

Levy, Marion J. (1952). The structure of society. Princeton, NJ: Princeton University Press.

Levy, Steven (2000). Insanely great: The life and times of Macintosh, the computer that changed everything. New York: Penguin.

Levy, Steven (2001). Hackers: Heroes of the computer revolution. New York: Penguin Books.

Lewis, Michael (2003). Moneyball: The art of winning an unfair game. New York: W. W. Norton.

Licklider, Joseph C. R., (1960). Man-Computer symbiosis. IRE Transactions on Human Factors in Electronics, 1(1): 4–11.

Lindtner, Silvia (2014). Hackerspaces and the Internet of Things in China: How makers are reinventing industrial production, innovation, and the self. China Information, 28(2): 145–167.

Lineweaver, Charles H., Fenner, Yeshe, & Gibson, Brad K. (2004). The galactic habitable zone and the age distribution of complex life in the Milky Way. Science, 303(5654): 59–62.

Linton, Ralph (1936). The study of man: An introduction. New York: Appleton-Century-Crofts.

Lisi, Antony G. (2015). I, for one, welcome our machine overlords. In John Brockman (Ed.). What to think about machines that think (pp. 22–24). New York: Harper Perennial.

Lissack, Michael R. (1999). Complexity: The science, its vocabulary, and its relation to organizations. Emergence, 1(1): 110–126.

Liu, Yang-Yu, Slotine, Jean-Jacques, & Barabási, Albert-László (2011). Controllability of complex networks. Nature, 473(7346): 167–173.

Livingstone, Sonia (2004). Media literacy and the challenge of new information and communication technologies. Communication Review, 7(1): 3–14.

Lloyd, David, Aon, Miguel A., & Cortassa, Sonia (2001). Why homeodynamics, not homeostasis? Scientific World Journal, 1: 133–145.

Loebbecke, Claudia, Bienert, Joerg, & Sunyaev, Ali (2013). A parallel platform for big data analytics: A design science approach. International Journal of Computer Science Engineering and Technology, 3(5): 152–156.

Loft, Anne (1995). Time is money. Culture and Organization, 1(1): 127–145.

Lohr, Steve (2012). The age of big data. http://www.nytimes.com/2012/02/12/sunday-review/big-datas-impact-in-the-world.html, last accessed 08 February 2016.

Lorenz, Edward N. (1963). Deterministic nonperiodic flow. Journal of the Atmospheric Sciences, 20(2): 130–141.

Louridas, Panagiotis (1999). Design as bricolage: Anthropology meets design thinking. Design Studies, 20(6): 17–535.

Lubatkin, Michael H., Simsek, Zeki, Ling, Yan, & Veiga, John F. (2006). Ambidexterity and performance in small-to medium-sized firms: The pivotal role of top management team behavioral integration. Journal of Management, 32(5): 646–672.

Luca, Michael, Kleinberg, Jon, & Mullainathan, Sendhil (2016). Algorithms need managers, too. Harvard Business Review, 94(1): 96–101.

Lucas, George (Director) (1997). Star Wars – A New Hope. San Rafael, CA: Lucasfilm.

Lucas, George (Director) (1980). Star Wars – The Empire Strikes Back. San Rafael, CA: Lucasfilm.

Luhmann, Niklas (1991). Soziale Systeme. Grundriss einer allgemeinen Theorie. Frankfurt am Main: Suhrkamp.

Luhmann, Niklas (2011). Einführung in die Systemtheorie. Heidelberg: Carl-Auer.

Lukoianova, Tatiana, & Rubin, Victoria L. (2014). Veracity roadmap: Is big data objective, truthful and credible? Advances in Classification Research Online, 24(1): 4–15.

Lupton, Deborah (2015). The thirteen Ps of big data. https://simplysociology.wordpress.com/2015/05/11/the-thirteen-ps-of-big-data, last accessed 03 February 2016.

Lycett, Mark (2013). Datafication: Making sense of (big) data in a complex world. European Journal of Information Systems, 22(4): 381–386.

Lynch, Clifford (2008). Big data: How do your data grow? Nature, 455(7209): 28–29.

Lyon, David (2003). Surveillance as social sorting: Privacy, risk, and digital discrimination. London: Routledge.

Machlup, Fritz (1962). Knowledge production and distribution in the United States. Princeton: Princeton University Press.

MacKenzie, Donald A. (1984). Marx and the machine. Technology and Culture, 25(3): 473–502.

MacKenzie, Donald A., & Wajcman, Judy (1985). The social shaping of technology. Berkshire: Open University Press.

MacQueen, James (1967). Some methods for classification and analysis of multivariate observations. Proceedings of the Fifth Berkeley Symposium on Mathematical Statistics and Probability, 1(14): 281–297.

Madrigal, Alexis C. (2013). IBM's Watson memorized the entire 'Urban Dictionary', then his overlords had to delete it. http://www.theatlantic.com/technology/archive/2013/01/ibms-watson-memorized-the-entire-urban-dictionary-then-his-overlords-had-to-delete-it/267047, last accessed 08 February 2016.

Mager, Astrid (2012). Algorithmic ideology: How capitalist society shapes search engines. Information, Communication & Society, 15(5): 769–787.

Maguire, Steve (2011) Constructing and appreciating complexity. In Peter Allen, Steve Maguire, & Bill McKelvey (Eds.). The SAGE handbook of complexity and management (pp. 79–92). Thousand Oaks: Sage.

Maguire, Steve, Allen, Peter, & McKelvey, Bill (2011) Complexity and management: Introducing the SAGE handbook. In Peter Allen, Steve Maguire, & Bill McKelvey (Eds.) The SAGE handbook of complexity and management (pp. 1–26). Thousand Oaks: Sage.

Mainzer, Klaus (2014). Die Berechnung der Welt. Von der Weltformel zu Big Data. München: C. H. Beck.

Mainzer, Klaus (2015). Industrie 4.0, richtig gestaltet, eröffnet neue Freiheitsgrade für die Menschen. G.I.B. Info, (4): 54–65.

Mandelbrot, Benoît B. (1997). Fractals: Form, chance and dimension, New York: W. H. Freeman.

Manovich, Lev (2011). Trending: The promises and challenges of big social data. In Matthew K. Cold (Ed.). Debates in the digital humanities (pp. 460–475). Minneapolis: University of Minnesota Press.

Mansell, Robin (2016). Power, hierarchy and the internet: why the internet empowers and disempowers. Global Studies Journal, 9(2): 19–25.

Manyika, James, Chui, Michael, Brown, Brad, Bughin, Jacques, Dobbs, Richard, Roxburgh, Charles, & Buyers, Angela H. (2011). Big data: The next frontier for innovation, competition and productivity. McKinsey Global Institute: 1–146.

March, James G. (1978). Bounded rationality, ambiguity, and the engineering of choice. Bell Journal of Economics, 9(2): 587–608.

March, James G. (1981). Footnotes to organizational change. Administrative Science Quarterly, 26(4): 563–577.

March, James G. (1991). Exploration and exploitation in organizational learning. Organization Science, 2(1): 71–87.

Marder, Eve (2015). Understanding brains: Details, intuition, and big data. PLoS Biology, 13(5): 1–6.

Marion, Russ (1999) The edge of organization: Chaos and complexity theories of formal social organizations. Thousand Oaks: Sage.

Marr, Bernard (2015). Is big data just a fad? http://www.weforum.org/agenda/2015/03/is-big-data-just-a-fad, last accessed 03 February 2016.

Marron, B. A., & de Maine, Paul A. D. (1967). Automatic data compression. Communications of the ACM, 10(11): 711–715.

Martin, Ben R. (2010). The origins of the concept of 'foresight' in science and technology: An insider's perspective. Technological Forecasting and Social Change, 77(9): 1438–1447.

Martin, Patricia Y., & Turner, Barry A. (1986). Grounded theory and organizational research. Journal of Applied Behavioural Science, 22(2): 141–157.

Marx, Karl (1971). The poverty of philosophy. New York: International Publishers.

Marx, Leo, & Smith, Merrit R. (1994). Introduction. In Merritt R. Smith, & Leo Marx (Eds.). Does technology drive history? The dilemma of technological determinism (pp. ix-xv). Cambridge, MA: MIT Press.

Marz, Nathan, & Warren, James (2005). Big data: Principles and best practices of scalable realtime data systems. Manning, NY: Manning Publications.

Mason, Daniel S., & Foster, William M. (2007). Putting Moneyball on ice? International Journal of Sport Finance, 2(4): 206–213.

Mass Effect 2 (2010). Edmonton: Bioware.

Maturana, Humberto R. (1970). Neurophysiology of cognition. In Paul L. Garvin (Ed.). Cognition. A multiple view (pp. 3–23). New York: Spartan Books.

Maturana, Humberto R. (1987). The biological foundation of self-consciousness and the physical domain of existence. In Eduardo R. Caianiello (Ed.) Physics of cognitive processes (pp. 324–379). Singapore: World Scientific.

Maturana, Humberto R., & Varela, Francisco J. (1972). Autopoiesis and cognition. The realization of the living. Dordrecht: Reidel Publishing.

Matzner, Tobias (2014). Why privacy is not enough privacy in the context of "ubiquitous computing" and "big data". Journal of Information, Communication and Ethics in Society, 12(2): 9–106.

Maxwell, James C. (1882). The life of James Clerk Maxwell. London: Macmillan.

Mayer-Schönberger, Viktor (2014). Neue Erkenntnisse über die Wirklichkeit. Forschung und Lehre, 21: 706–707.

Mayer-Schönberger, Viktor, & Cukier, Kenneth (2013). Big data: A revolution that will transform how we live, work, and think. Boston: Houghton Mifflin Harcourt.

Mayer, Marta C., & Pirri, Fiora (1996). Abduction is not deduction-in-reverse. Logic Journal of IGPL, 4(1): 95–108.

McAfee, Andrew, & Brynjolfsson, Eric (2012). Big data: The management revolution. Harvard Business Review, 90(10): 60–68.

McCandless, David (2010). Data, information, knowledge, wisdom, http://www.informationisbeautiful.net/2010/data-information-knowledge-wisdom, last accessed 03 February 2016).

McCann, Philip (2008). Globalization and economic geography: The world is curved, not flat. Cambridge Journal of Regions, Economy and Society, 1(3): 351–370.

McCarthy, John (2007). From here to human-level AI. Artificial Intelligence, 171(18): 1174–1182.

McCorduck, Pamela (2015). An epochal human event. In John Brockman (Ed.). What to think about machines that think (pp. 51–53). New York: Harper Perennial.

McDonald, Robert (2011). Inside P&G's digital revolution. McKinsey Quarterly, http://www.mckinsey.com/industries/consumer-packaged-goods/our-insights/inside-p-and-ampgs-digital-revolution, last accessed 25 April 2016.

McFedries, Paul (2013). Tracking the quantified self [Technically speaking]. Spectrum, IEEE, 50(8): 24–24.

McKelvey, Bill (2004). Toward a complexity science of entrepreneurship. Journal of Business Venturing, 19(3): 313–341.

McKelvey, Bill (2016). Complexity Ingredients Required for Entrepreneurial Success. Entrepreneurship Research Journal, 6(1): 53–73.

McLain, David L., & Keenan, John P. (1999). Risk, information, and the decision about response to wrongdoing in an organization. Journal of Business Ethics, 19(3): 255–271.

McLeod, Kari S. (2000). Our sense of Snow: The myth of John Snow in medical geography. Social Science & Medicine, 50(7): 923–935.

McLuhan, Marshall (1967). The medium is the massage: An inventory of effects. London: Penguin Books.

McNeely, Connie L., & Hahm, Jong-On (2014). The big (data) bang: Policy, prospects, and challenges. Review of Policy Research, 31(4): 304–310.

Mead, George H. (1934). Mind, self, and society. Chicago: University of Chicago Press.

Medina, Eden (2006). Designing freedom, regulating a nation: Socialist cybernetics in Allende's Chile. Journal of Latin American Studies, 38(3): 571–606.

Meehl, Paul E., & Rosen, Albert (1955). Antecedent probability and the efficiency of psychometric signs, patterns, or cutting scores. Psychological Bulletin, 52(3): 194–216.

Meijs, Michel (2002). The myth of manageability of corporate identity. Corporate Reputation Review, 5(1): 20–34.

Merali, Yasmin, & Allen, Peter (2011). Complexity and systems thinking. In Peter Allen, Steve Maguire, & Bill McKelvey (Eds.). The SAGE handbook of complexity and management (pp. 31–52). Thousand Oaks: Sage.

Merriam, Daniel F. (1974). Resource and environmental data analysis. Geological Survey Professional Paper, 921: 37–45.

Merton, Robert K. (1936). The unanticipated consequences of purposive social action. American Sociological Review, 1(6): 894–904.

Merton, Robert K. (1948). The self-fulfilling prophecy. The Antioch Review, 8(2):193–210.

Merton, Robert K. (1961). Singletons and multiples in scientific discovery: A chapter in the sociology of science. Proceedings of the American Philosophical Society, 105(5): 470–486.

Metcalf, Jacob, & Crawford, Kate (2016). Where are human subjects in big data research? The emerging ethics divide. Big Data & Society, 3, published online before print.

Miceli, Marcia P., & Near, Janet P. (1994). Whistleblowing: Reaping the benefits. Academy of Management Executive, 8(3): 65–72.

Michael, Katina, & Michael, M. G. (2013). No limits to watching. Communications of the ACM, 56(11): 26–28.

Microsoft (2013). The big bang: How the big data explosion is changing the world. https://news.microsoft.com/2013/02/11/the-big-bang-how-the-big-data-explosion-is-changing-the-world/, last accessed 03 February 2016.

Miller, Arthur R. (1971). The assault on privacy. Computers, data banks, and dossier. Ann Arbor: University of Michigan Press.

Miller, Claire C. (2015). Can an algorithm hire better than a human? http://www.nytimes.com/2015/06/26/upshot/can-an-algorithm-hire-better-than-a-human.html?_r=0, last accessed 04 February 2016.

Miller, Gregory S. (2006). The press as a watchdog for accounting fraud. Journal of Accounting Research, 44(5): 1001–1033.

Miller, H. Gilbert, & Mork, Peter (2013). From data to decisions: A value chain for big data. IT Professional, 15(1): 57–59.

Miller, Harvey J. (2010). The data avalanche is here. Shouldn't we be digging? Journal of Regional Science, 50(1): 181–201.

Miller, John H. (2015). A crude look at the whole. The science of complex systems in business, life, and society. New York: Basic Books.

Miller, Kent D. (2008). Simon and Polanyi on rationality and knowledge. Organization Studies, 29(7): 933–955.

Miller, Marc (2013). HR and big data: Not yet, first things first! Workforce Solutions Review, August/September: 39–40.

Miller, Richard W. (1984). Analyzing Marx: Morality, power and history. Princeton, NJ: Princeton University Press.

Minecraft (2009). Stockholm: Mojang.

Minsky, Marvin (1961). Steps toward artificial intelligence. Proceedings of the IRE, 49(1): 8–30.

Mintzberg, Henry, & McHugh, Alexandra (1985). Strategy formation in an adhocracy. Administrative Science Quarterly 30(2): 160–197.

Mitchell, Tom M. (1997) Machine learning. New York: McGraw-Hill.

Mittal, Sparsh (2016). A survey of techniques for approximate computing. ACM Computing Surveys, 48(4): 62.

Mittelstadt, Brent D., & Floridi, Luciano (2015). The ethics of big data: current and foreseeable issues in biomedical contexts. Science and Engineering Ethics, published online before print: 1–39.

Mohri, Mehryar, Rostamizadeh, Afshin, & Talwalkar, Ameet (2012). Foundations of machine learning. Cambridge: MIT Press.

Monroe, Burt L., Pan, Jennifer, Roberts, Margaret E., Sen, Maya, & Sinclair, Betsy (2015). No! Formal theory, causal inference, and big data are not contradictory trends in political science. PS: Political Science & Politics, 48(1): 71–74.

Moore, Don A. & Healy, Paul J. (2008). The trouble with overconfidence. Psychological Review, 115(2): 502–517.

Mordvintsev, Alexander, Olah, Christopher, & Tyka, Mike (2015). Inceptionism: Going deeper into neural networks, http://googleresearch.blogspot.co.uk/2015/06/inceptionism-going-deeper-into-neural.html, last accessed 08 February 2016.

Moretti, Franco (2013). Distant reading. London: Verso.

Morgan, Gareth (1982). Cybernetics and organization theory: Epistemology or technique? Human Relations, 35(7): 521–537.

Morin, Edgar (2008). On complexity. Cresskill: Hampton Press.

Morin, Edgar, & Coppay, Frank (1983). Beyond determinism: The dialogue of order and disorder. SubStance, 12(3): 22–35.

Morozov, Evgeny (2013). To save everything, click here: Technology, solutionism, and the urge to fix problems that don't exist. Philadelphia: Perseus Books.

Morris, Robert J. T., & Truskowski, Brian J. (2003). The evolution of storage systems. IBM systems Journal, 42(4): 205–217.

Morrison, Keith (2005). Structuration theory, habitus and complexity theory: Elective affinities or old wine in new bottles? British Journal of Sociology of Education, 26(3): 311–326.

Müller, Paul J. (1979). Data protection and social research – International perspectives. In Ekkehard Mochmann, & Paul J. Müller (Eds.). Data protection and social science research (pp. 11–26). Frankfurt am Main: Campus.

Munakata, Toshinori (1998). Fundamentals of the new artificial intelligence. Heidelberg: Springer.

Murdoch, Jonathan (1997). Inhuman/nonhuman/human: Actor-network theory and the prospects for a nondualistic and symmetrical perspective on nature and society. Environment and Planning D: Society and Space, 15(6): 731–756.

Murthy, Dhiraj, & Bowman, Saqyer A. (2014). Big data solutions on a small scale: Evaluating accessible high-performance computing for social research. Big Data & Society, 1(2): 1–12.

Nafus, Dawn, & Sherman, Jamie (2014). This one does not go up to 11: The quantified self movement as an alternative big data practice. International Journal of Communication, 8: 1784–1794.

Narayanan, Arvind, & Shmatikov, Vitaly (2008). Robust de-anonymization of large sparse datasets. IEEE Symposium on Security and Privacy, Oakland, CA, 111–125.

Narayanan, Arvind, & Shmatikov, Vitaly (2010). Myths and fallacies of personally identifiable information. Communications of the ACM, 53(6): 24–26.

New Scientist (2015). Rosetta. Touching down on a comet. https://newscientist.atavist.com/rosetta, last accessed 25 April 2016.

Newell, Sue, & Marabelli, Marco (2015). Strategic opportunities (and challenges) of algorithmic decision-making: A call for action on the long-term societal effects of 'datification'. Journal of Strategic Information Systems, 24(1): 3–14.

Newman, Abraham. L. (2015b). What the 'right to be forgotten' means for privacy in a digital age, Science, 347(6221): 507–508.

Newman, Blair (2015a). The pioneering AC Milan lab. https://thesefootballtimes.co/2015/01/15/the-ac-milan-lab, last accessed 25 April 2016.

Newsom, S. W. B. (2006). Pioneers in infection control: John Snow, Henry Whitehead, the Broad Street pump, and the beginnings of geographical epidemiology. Journal of Hospital Infection, 64(3): 210–216.

Nicholson, Scott (2012). A user-centered theoretical framework for meaningful gamification. Proceedings of Games+Learning+Society 8.0, Madison, WI, 223–230.

Nickerson, David W., & Rogers, Todd (2014). Political campaigns and big data. Journal of Economic Perspectives, 28(2): 51–73.

Nicolis, Gregoire, & Prigogine, Ilya (1977). Self-organization in nonequilibrium systems. Hoboken, NJ: Wiley.

Nilsson, Nils J. (2005). Human-level artificial intelligence? Be serious! AI Magazine, 26(4): 68–75.

Nisen, Max (2014). Google came up with a formula for deciding who gets promoted – here's what happened. http://qz.com/299112/google-came-up-with-a-formula-for-deciding-who-gets-promoted-heres-what-happened, last accessed 25 April 2016.

Nissenbaum, Helen (1996). Accountability in a computerized society. Science and Engineering Ethics, 2(1): 25–42.

NIST Big Data Public Working Group (2014). Big data interoperability framework: Definitions. Gaithersburg, MD: National Institute of Standards and Technology.

Nolan, Rachel (2012). Behind the cover story: How much does Target know? http://6thfloor.blogs.nytimes.com/2012/02/21/behind-the-cover-story-how-much-does-target-know, last accessed 16 February 2016.

Nonaka, Ikujiro, & Takeuchi, Hirotaka (1995). The knowledge-creating company: How Japanese companies create the dynamics of innovation. Oxford: Oxford University Press.

O'Leary, Daniel E. (2013). Artificial intelligence and big data. IEEE Intelligent Systems, 28(2): 96–99.

O'Hara, Kieron, & Shadbolt, Nigel (2008). The spy in the coffee machine. The end of privacy as we know it. London: Oneworld.

O'Neil, Cathy (2012). Let them game the model, http://mathbabe.org/2012/02/03/let-them-game-the-model, last accessed 26 April 2016.

Obbema, Fokke, Vlaskamp, Marije, & Persson, Michael (2015). China rates its own citizens – including online behavior. http://www.volkskrant.nl/buitenland/china-rates-its-own-citizens-including-online-behaviour~a3979668, last accessed 25 April 2016.

Ohm, Paul (2010). Broken promises of privacy: Responding to the surprising failure of anonymization. UCLA Law Review, 57: 1701–1777.

Olavsrud, Thor (2014). How big data is helping to save the planet. http://www.cio.com/article/2683133/big-data/how-big-data-is-helping-to-save-the-planet.html, last accessed 09 February 2016.

Olejnik, Lukasz, Acar, Gunes, Castelluccia, Claude, & Diaz, Claudia (2015) The leaking battery: A privacy analysis of the HTML5 Battery Status API. Cryptology ePrint Archive, Report 2015/616: 1–9.

Oliver, Dean (2004). Basketball on paper: Rules and tools for performance analysis. Washington, D.C.: Brassey's.

Olson, Donald R., Konty, Kevin J., Paladini, Marc, Viboud, Cecile, Simonsen, Lone (2013) Reassessing Google flu trends data for detection of seasonal and pandemic influenza: A comparative epidemiological study at three geographic scales. PLoS Computational Biology, 9(10): 1–11.

Oprescu, Florin, Jones, Christian, Katsikitis, Mary (2014) I play at work – Ten principles for transforming work processes through gamification. Frontiers in Psychology, 5: 1–5.

Orbach, Maya, Demko, Maegen, Doyle, Jeremy, Waber, Ben N., & Pentland, A. Sandy (2015). Sensing informal networks in organizations. American Behavioral Scientist, 59(4): 508–524.

Orlikowski, Wanda J. (1992). The duality of technology: Rethinking the concept of technology in organizations. Organization Science, 3(3): 398–427.

Orwell, George (1949). Nineteen eighty-four. Secker and Warburg, London.

Oster, Gary (2009). Building innovation capacity in emerging markets. Effective Executive, 13(1): 10–16.

Oswald, Margit E., & Grosjean, Stefan (2004). Confirmation bias. In Rüdiger F. Pohl (Ed.). Cognitive illusions: A handbook on fallacies and biases in thinking, judgement and memory (pp. 79–96). Hove and New York: Psychology Press.

Otto, Paul N., Antón, Annie I., & Baumer, David L. (2007). The ChoicePoint dilemma: How data brokers should handle the privacy of personal information. IEEE Security & Privacy, 5(5): 15–23.

Özdemir, Vural, Badr, Kamal F., Dove, Edward S., Endrenyi, Laszlo, Geraci, Christy J., Hotez, Peter J., Milius, Djims, Neves-Pereira, Maria, Pang, Tikki, Rotimi, Charles, N., Sabra, Ramzi, Sarkissian, Chrtineh N., Srivastava, Sanjeeva, Tims, Hesther, Zgheib, Nathalie K., & Kickbusch, Ilona (2013). Crowd-funded micro-grants for genomics and "big data": An actionable idea connecting small (artisan) science, infrastructure science and citizen philantrophy. Omics: A Journal of Integrative Biology, 17(4): 161–172.

Palmas, Karl (2011). Predicting what you'll do tomorrow: Panspectric surveillance and the contemporary corporation. Surveillance & Society, 8(3): 338–354.

Pannabecker, John R. (1991). Technological impacts and determinism in technology education: Alternate metaphors from social constructivism. Journal of Technology Education, 3(1), http://scholar.lib.vt.edu/ejournals/JTE/v3n1/html/pannabecker.html, last accessed 28 April 2016.

Pantzar, Mika, & Shove, Elizabeth (Eds.) (2005) Manufacturing leisure. Helsinki: National Consumer Research Centre.

Parameswaran, Ashwin (2013). How to commit fraud and get away with it: A guide for CEOs. http://www.macroresilience.com/2013/12/04/how-to-commit-fraud-and-get-away-with-it-a-guide-for-ceos/, last accessed 23 April 2016.

Parenti, Christian (2001). Big brother's corporate cousin: High-tech workplace surveillance is the hallmark of a new digital Taylorism. The Nation, 273(5): 26–30.

Pariser, Eli (2011). The filter bubble: What the Internet is hiding from you. Penguin: London.

Park, Sungmee, & Jayaraman, Sundaresan (2003). Enhancing the quality of life through wearable technology. IEEE Engineering in Medicine and Biology Magazine, 22(3): 41–48.

Parker, Martin (1998). Judgement day: Cyberorganization, humanism and postmodern ethics. Organization, 5(4): 503–518.

Parry, Emma (2014). E-HRM: A catalyst for changing the HR function? In Francisco J. Martínez-López (Ed.). Handbook of strategic e-Business management (pp. 589–604). Heidelberg: Springer.

Parsons, Talcott (1951). Illness and the role of the physician: A sociological perspective. American Journal of Orthopsychiatry, 21(3): 452–460.

Pasquale, Frank (2015). The black box society. The secret algorithms that control money and information. Cambridge: Harvard University Press.

Patrick, Robert L. (1977). Performance assurance and data integrity practices. Gaithersburg, MD: National Bureau of Standards.

Paul, Christopher A. (2011). Optimizing play: How theorycraft changes gameplay and design. Game Studies, 11(2), http://gamestudies.org/1102/articles/paul, last accessed 28 April 2016.

Pawson, Ray, Wong, Geoff, & Owen, Lesley (2011). Known knowns, known unknowns, unknown unknowns: the predicament of evidence-based policy. American Journal of Evaluation, 32(4): 518–546.

Peck, Don (2013). They're watching you at work. http://www.theatlantic.com/magazine/archive/2013/12/theyre-watching-you-at-work/354681, last accessed 10 June 2016.

Peirce, Charles S. (1958). Collected works: 1931–1958. Cambridge: Harvard University Press.

Pentland, A. Sandy (2010). To signal is human. Real-time data mining unmasks the power of imitation, kith and charisma in our face-to-face social networks, American Scientist, 98(3): 204–211.

Pentland, A. Sandy (2014). Social physics: How good ideas spread – The lessons from a new science. New York: Penguin Books.

Pescosolido, Bernice A. (1992). Beyond rational choice: The social dynamics of how people seek help. American Journal of Sociology, 97(4): 1096–1138.

Peteranderl, Sonja (2016). Notruf eins null null null eins null. Wired, 2(1): 13–15.

Peters, Brad (2012). The big data gold rush. http://www.forbes.com/sites/bradpeters/2012/06/21/the-big-data-gold-rush/#6b87676a5710, last accessed 26 April 2016.

Pickles, John (Ed.) (1995). Ground truth: The social implications of geographic information systems. New York: Guilford Press.

Pierre, Magali, Jemelin, Christophe, & Louvet, Nicolas (2011). Driving an electric vehicle. A sociological analysis on pioneer users. Energy Efficiency, 4(4): 511–522.

Pinch, Trevor J. (2009). The social construction of technology (SCOT): The old, the new, and the nonhuman. In Phillip Vannini (Ed.). Material culture and technology in everyday life: Ethnographic approaches (pp. 45–58). New York: Peter Lang.

Pinch, Trevor J., & Bijker, Wiebe E. (1984). The social construction of facts and artefacts: Or how the sociology of science and the sociology of technology might benefit each other. Social Studies of Science, 14(3): 399–441.

Pittinsky, Todd L. (2016). We're making the wrong case for diversity in Silicon Valley. https://hbr.org/2016/04/were-making-the-wrong-case-for-diversity-in-silicon-valley, last accessed 26 April 2016.

Poincaré, Henri (1881). Mémoire sur les courbes définies par une équation différentielle. Journal de Mathématiques Pures et Appliquées, 3(7): 375–422.

Poincaré, Henri (1914). Science and method. London: Thomas Nelson and Sons.

Pongratz, Hans J., & Voß, G. Günter (1997). Fremdorganisierte Selbstorganisation. Eine soziologische Diskussion aktueller Managementkonzepte. Zeitschrift für Personalforschung, 11(1): 30–53.

Poole, David, Mackworth, Alan, & Goebel, Randy (1998). Computational intelligence: A logical approach. Oxford: Oxford University Press.

Popper, Karl R. (1959). The logic of scientific discovery. London: Routledge.

Popper, Karl. R. (1963). Conjectures and refutations. London: Routledge.

Porter, Theodore M. (1996). Trust in numbers: The pursuit of objectivity in science and public life. Princeton: Princeton University Press.

Postman, Neil (1992). Technopoly. New York: Vintage Books.

Pouvreau, David, & Drack, Manfred (2007). On the history of Ludwig von Bertalanffy's "General Systemology", and on its relationship to cybernetics: Part I: Elements on the origins and genesis of Ludwig von Bertalanffy's "General Systemology". International Journal of General Systems, 36(3): 281–337.

Pratchett, Terry, & Baxter, Stephen (2012). The long earth. New York: Doubleday.

Prensky, Marc (2001). Digital natives, digital immigrants. On the Horizon, 9(5): 1–6.

Prensky, Marc (2009). H. sapiens digital: From digital immigrants and digital natives to digital wisdom. Innovate: Journal of Online Education, 5(3), http://nsuworks.

nova.edu/cgi/viewcontent.cgi?article=1020&context=innovate, last accessed 09 February 2016.

Preskill, John (1998). Quantum computing: Pro and con. Proceedings of the Royal Society of London A: Mathematical, Physical and Engineering Sciences, 454(1969): 469–486.

Priestly, Jennifer (2015). We're all data geeks now, IEEE Spectrum, 52(8): 29.

Prigogine, Ilya, & Stengers, Isabelle (1984). Order out of chaos. New York: Bantam Books.

Pronin, Emily, Lin, Daniel Y., & Ross, Lee (2002). The bias blind spot: Perceptions of bias in self versus others. Personality and Social Psychology Bulletin, 28(3): 369–381.

Provine, William B. (1986). Sewall Wright and evolutionary biology. Chicago: University of Chicago Press.

Provost, Foster, & Fawcett, Tom (2013). Data science and its relationship to big data and data-driven decision making. Big Data, 1(1): 51–59.

Puschmann, Cornelius, & Burgess, Jean (2014). Metaphors of big data. International Journal of Communication, 8: 1690–1709.

Qiu, Jack L. (2015). Reflections on Big Data: 'Just because it is accessible does not make it ethical'. Media, Culture & Society, 37(7): 1089–1094.

Quan-Haase, Anabel (2016). Technology & Society. Social networks, power, and inequality. Don Mills, Ontario: Oxford University Press.

Raftopoulos, Marigo (2014). Towards gamification transparency: A conceptual framework for the development of responsible gamified enterprise systems. Journal of Gaming & Virtual Worlds, 6(2): 159–178.

Ramo, Danielle E., Rodriguez, Theresa M., Chavez, Kathryn, Sommer, Markus J., & Prochaska, Judith J. (2014). Facebook recruitment of young adult smokers for a cessation trial: methods, metrics, and lessons learned. Internet Interventions, 1(2): 58–64.

Rao, Hayagreeva (1998). Caveat emptor: The construction of nonprofit consumer watchdog organizations. American Journal of Sociology, 103(4): 912–961.

Rätsch, Christian (2015). Big data führt zum Stillstand. http://www.manager-magazin.de/unternehmen/it/big-data-verhindert-innovationen-in-unternehmen-a-1034854.html, last accessed 08 March 2016.

Rawls, John (1971). A theory of justice. Cambridge: Harvard University Press.

Rayport, Jeffrey F. (2011). What big data needs: A code of ethical practices. http://www.technologyreview.com/news/424104/what-big-data-needs-a-code-of-ethical-practices, last accessed 24 April 2016.

Reay, Diane (2004). 'It's all becoming a habitus': Beyond the habitual use of habitus in educational research. British Journal of Sociology of Education, 25(4): 431–444.

Reeves, Martin, Zeng, Ming, & Venjara, Amin (2015) The self-tuning enterprise. Harvard Business Review, 93(6): 76–83.

Renn, Ortwin (2008). Risk governance: Coping with uncertainty in a complex world. London: Earthscan.

Renn, Ortwin, Klinke, Andreas, van Asselt, Marjolein B. A. (2011) Coping with complexity, uncertainty and risk governance: A synthesis. Ambio, 40(2): 231–246.

Rensberger, Boyce (2009). Science journalism: Too close for comfort. Nature, 459(7250): 1055–1056.

Richards, Neil M., & King, Jonathan H. (2013). Three paradoxes of big data. Stanford Law Review Online, 66: 41–46.

Richardson, James H. (2014). The Spotify paradox: How the creation of a compulsory license scheme for streaming on-demand music platforms can save the music industry. UCLA Entertainment Law Review, 22: 45–95.

Richtel, Matt (2013). How big data is playing recruiter for specialized workers. http://www.nytimes.com/2013/04/28/technology/how-big-data-is-playing-recruiter-for-specialized-workers.html, last accessed 28 April 2016.

Rieley, James B. (2000). Are your employees gaming the system? National Productivity Review, 19(3): 1–6.

Rindova, Violina P., & Kotha, Suresh (2001). Continuous "morphing": Competing through dynamic capabilities, form, and function. Academy of Management Journal, 44(6): 1263–1280.

Robins, Richard W., Spranca, Mark D., & Mendelsohn, Gerald A. (1996). The actor-observer effect revisited: Effects of individual differences and repeated social interactions on actor and observer attributions. Journal of Personality and Social Psychology, 71(2): 375–389.

Roncallo-Dow, Sergio, Uribe-Jongbloed, Enrique, Barker, Kim, & Scholz, Tobias M. (2013). Authorship in virtual worlds: Author's death to rights revival? Journal for Virtual Worlds Research, 6(3): 1–15.

Rosen, Jeffrey (2012). The right to be forgotten. Stanford Law Review Online, 64: 88–92.

Rosenberg, Daniel (2013). Data before the fact. In Lisa Gitelman (Ed.). "Raw data" is an oxymoron (pp. 15–40). Cambridge: MIT Press.

Ross, Lee, & Ward, Andrew (1997). Naive realism in everyday life: Implications for social conflict and misunderstanding. In Edward S. Reed, Elliot Turiel, & Terrance Brown (Eds.). Values and knowledge (pp. 103–135). Mahwah, NJ: Lawrence Erlbaum.

Rosser, James C., Lynch, Paul J., Cuddihy, Laurie, Gentile, Douglas A., Klonsky, Jonathan, & Merrell, Ronald (2007). The impact of video games on training surgeons in the 21st century. Archives of Surgery, 142(2): 181–186.

Rousseau, Denise M., Sitkin, Sim B., Burt, Ronald S., & Camerer, Colin (1998). Not so different after all: A cross-discipline view of trust. Academy of Management Review, 23(3): 393–404.

Rousseau, Denise M., & Tijoriwala, Snehal A. (1998). Assessing psychological contracts: Issues, alternatives and measures. Journal of Organizational Behavior, 19(1): 679–695.

Rubinstein, Ira S. (2013). Big data: the end of privacy or a new beginning? International Data Privacy Law, 3(2): 74–87.

Ruckenstein, Minna, & Pantzar, Mika (2015). Beyond the quantified self: Thematic exploration of a dataistic paradigm. New Media & Society, published online before print.

Rumelhart, David E., Hinton, Geoffrey E., & Williams, Ronald J. (1986). Learning representations by back-propagating errors. Nature, 323(6088): 533–536.

Runkel, Phillip J., & Runkel, Margaret (1984). A guide to usage for writers and students in the social sciences. Totowa, NJ: Rowman and Allanheld.

Russel, Stuart J. & Norvig, Peter (1995). Artificial intelligence: A modern approach. Englewood Cliffs, NJ: Prentice Hall.

Russell, Chuck, & Bennett, Nathan (2014). Big data and talent management: Using hard data to make the soft stuff easy. Business Horizons, 58(3): 237–242.

Russom, Philip. (2013). Managing big data. TDWI Best Practices Report, 1–40.

Ruths, Derek, & Pfeffer, Jürgen (2014). Social media for large studies of behavior. Science 346(6213): 1063–1064.

Ryan, Ann M., & Ployhart, Robert E. (2014). A century of selection. Annual Review of Psychology, 65: 693–717.

Ryan, Richard M., & Deci, Edward L. (2000). Self-determination theory and the facilitation of intrinsic motivation, social development, and well-being. American Psychologist, 55(1): 68–78.

Sagan, Carl (1996). The demon-haunted world. Science as a candle in the dark. New York: Ballantine Books.

Sagie, Abraham, Elizur, Dov, & Koslowsky, Meni (1990). Effect of participation in strategic and tactical decisions on acceptance of planned change. The Journal of Social Psychology, 130(4): 459–465.

Sahay, Sundeep (2003). Global software alliances: the challenge of 'standardization'. Scandinavian Journal of Information Systems, 15(1): 3–21.

Salen, Katie, & Zimmerman, Eric (2004). Rules of play. Game design fundamentals. Cambridge: MIT Press.

Samuels, Warren J. (2000). Signs, pragmatism, and abduction: The tragedy, irony, and promise of Charles Sanders Peirce. Journal of Economic Issues, 34(1): 207–217.

Sanchez, Ron, Heene, Aimé, & Thomas, Howard (1996). Introduction: Towards the theory and practice of competence-based competition. In Ron Sanchez, Aimé Heene, & Howard Thomas (Eds.). Dynamics of competence-based competition. Theory and practice in the new strategic management (pp. 1–35). Oxford: Pergamon.

Saqib, Sheikh M., Khan, Hamid, M., Mahmood, K., & Naeem, T. (2015). BIG-data challenges: A review on existing solutions. American Journal of Information Science and Computer Engineering, 1(2): 38–43.

Sawyer, R. Keith (2002). Durkheim's dilemma: Toward a sociology of emergence. Sociological Theory, 20(2): 227–247.

Schaller, Robert R. (1997) Moore's law: Past, present and future. IEEE Spectrum, 34(6): 52–59.

Schank, Roger (2015). Machines that think are in the movies. In John Brockman (Ed.). What to think about machines that think (pp. 132–135). New York: Harper Perennial.

Schneier, Bruce (2015). Data and goliath. The hidden battle to collect your data and control your world. New York: W. W. Norton.

Schoen, Harald, Gayo-Avello, Daniel, Takis Metaxas, Panagiotis, Mustafaraj, Eni, Strohmaier, Markus, & Gloor, Peter (2013). The power of prediction with social media. Internet Research, 23(5): 528–543.

Scholz, Christian (1984). OR/MS methodology – A conceptual framework, Omega, 12(1): 53–61.

Scholz, Christian, & Josephy, Norman (1984). Industry analysis: The pattern approach. Harvard Business School Working Paper, 84–44.

Scholz, Christian (1987). Corporate culture and strategy – The problem of strategic fit. Long Range Planning, 20(4): 78–87.

Scholz, Christian (2000). Strategische Organisation. Multiperspektivität und Virtualität. Landsberg, Lech: moderne industrie.

Scholz, Christian (2014a). Personalmanagement. Informationsorientierte und verhaltenstheoretische Grundlagen. München: Vahlen.

Scholz, Christian (2014b). Big Data in der Personalarbeit: Nein, danke, http://derstandard.at/1388650688167/Big-Data-in-der-Personalarbeit-Nein-danke, last accessed 26 April 2016.

Scholz, Christian (2016). Smart durch Big Data? Plädoyer für eine kritische Betrachtung. Wirtschaftspsychologie Aktuell, 16 (2/2016), 25–30.

Scholz, Tobias M. (2012). Talent management in the video game industry: The role of cultural diversity and cultural intelligence. Thunderbird International Business Review, 54(6): 845–858.

Scholz, Tobias M. (2013a). Complex systems in organizations and their influence on human resource management. In Thomas Gilbert, Markus Kirkilionis, & Gregoire

Nicolis (Eds.). Proceedings of the European conference on complex systems (pp. 745–750). Heidelberg: Springer.

Scholz, Tobias M. (2013b). Does context matter? Conceptualizing relational contextualization. In Konstantin Mitgutsch, Simon Huber, Jeffrey Wimmer, Mcihael G. Wagner, & Herbert Rosenstingl (Eds.). Context matters! Exploring and reframing games and play in context (pp. 89–98). Vienna: new academic press.

Scholz, Tobias M. (2013c). Spielend arbeiten – Parallelen zwischen der "World of Warcraft" und der "World of Workcraft". In Bundesministerium für Wirtschaft, Familie und Jugend (Ed.). Game Over. Was nun? Vom Nutzen und Nachteil des digitalen Spiels für das Leben (pp. 107–117). Vienna: Bundesministerium für Wirtschaft, Familie und Jugend.

Scholz, Tobias M. (2014). Data in faculties – the dean's role in the brave new (data) world. In Christian Scholz, & Volker Stein (Eds.). The dean in the university of the future (pp. 155–161). München-Mering: Hampp.

Scholz, Tobias M. (2015a). The impact of big data on the organization from an evolutionary perspective. 31th European Group for Organizational Studies Colloquium (EGOS), 1–21.

Scholz, Tobias M. (2015b). The human role within organizational change: A complex system perspective. In Frank E. P. Dievernich, Kim O. Tokarski, & Jie Gong (Eds.). Change management and the human factor: Advances, Challenges and Contradictions in Organizational Development (pp. 19–31). Heidelberg: Springer.

Scholz, Tobias M. (2015c). Game leadership – What can we learn from competitive games? In Julia Hiltscher, & Tobias M. Scholz (Eds.). eSports yearbook 2013/14 (pp. 93–106). Norderstedt: Books on Demand.

Scholz, Tobias M. (2016a). Una mirada a la textura causal e identidades múltiples para entender a los gurmés digitales – Una observación teórica. In Sergio Roncallo-Dow, Enrique Uribe-Jongbloed, Eduardo Gutiérrez (Eds.) Identidades, Héroes y discursos en la modernidad tardía (pp. 151–163). Chía, Colombia: Universidad de La Sabana Collección Compilaciones.

Scholz, Tobias M. (2016b). Language as means of dynamizing organizations. 32th European Group for Organizational Studies Colloquium (EGOS), 1–8.

Scholz, Tobias M., & Reichstein, Matthis S. (2015): Wenn neue Paradigmen in die Gestaltung von Arbeitswelten eingreifen: Hacker-Ethos in der Digitalisierung. In Stephan Habscheid, Gero Hoch, Hilde Schröteler-von Brandt, & Volker Stein (Eds.). Zum Thema: Gestalten gestalten. Diagonal Heft 36 (pp. 135–148). Göttingen: Vandenhoeck & Ruprecht.

Schramm-Klein, Hanna, Zentes, Joachim, Steinmann, Sascha, Swoboda, Bernhard, & Morschett, Dirk (2016). Retailer corporate social responsibility is relevant to consumer behavior. Business & Society, 55(4): 550–575.

Schroeck, Michael, Shockley, Rebecca, Smart, Janet, Romero-Morales, Dolores, & Tufano, Peter (2012). Analytics: The real-world use of big data – How innovative enterprises extract value from uncertain data. IBM Global Business Services, 1–20.

Schroeder, Ralph (2014). Big data and the brave new world of social media research. Big Data & Society, 1(2): 1–11.

Schulz, Kathryn (2011). What is distant reading? http://www.nytimes.com/2011/ 06/26/ books/review/the-mechanic-muse-what-is-distant-reading.html, last accessed 28 April 2016.

Schumpeter, Joseph A. (1942). Capitalism, socialism, & democracy. London: Routledge.

Schwefel, Hans Paul (1994). On the evolution of evolutionary computation. Proceedings of 3rd International Conference of IEEE World Congress on Computer Intelligence, Orlando, FL, 116–124.

Searle, John R. (1980). Minds, brains, and programs. Behavioral and Brain Sciences, 3(3): 417–424.

Seaver, Nick (2015). The nice thing about context is that everyone has it. Media, Culture & Society, 37(7): 1101–1109.

Segal, Howard P. (1985). Technological utopianism in American culture. Chicago: University of Chicago Press.

Seife, Charles (2015). Big data: The revolution is digitized. Nature, 518(7540): 480–481.

Sellen, Abigail, Rogers, Yvonne, Harper, Richard, & Rodden, Tom (2009). Reflecting human values in the digital age. Communications of the ACM, 52(3): 58–66.

Seltzer, William (2005). The promise and pitfalls of data mining: Ethical issues. Proceedings of the American Statistical Association, Section on Government Statistics, 1441–1445.

Selvin, Hanan C., & Stuart, Alan (1966). Data-dredging procedures in survey analysis. The American Statistician, 20(3): 20–23.

Senge, Peter (1990). The fifth discipline. The art and science of the learning organization. New York: Currency Doubleday.

Servick, Kelly (2015). Proposed study would closely track 10,000 New Yorkers. Science, 350(6260): 493–494.

Shah, Shvetank, Horne, Andrew, & Capellá, Jaime (2012). Good data won't guarantee good decisions. Harvard Business Review, 90(4): 23–25.

Shaw, Adrienne (2010). What Is video game culture? Cultural studies and game studies. Games and Culture, 5(4): 403–424.

Shaw, Jonathan (2014). Why 'big data' is a big deal, Harvard Magazine, (March-April): 30–36.

Shein, Esther (2014): Should everybody learn to code? Communications of the ACM 57(2): 16–18.

Shen, Yung-Cheng, Bei, Lien-Ti, & Chu, Chia-Hsien (2011). Consumer evaluations of brand extension: The roles of case-based reminding on brand-to-brand similarity. Psychology & Marketing 28(1): 91–113.

Shepard, Lea (2013). Seeking solutions to financial history discrimination. Connecticut Law Review, 46(3): 993–1044.

Shepherd, Dean A., & Sutcliffe, Kathleen M. (2011). Inductive top-down theorizing: A source of new theories of organization. Academy of Management Review, 36(2): 361–380.

Shermer, Michael (2008). Patternicity: Finding meaningful patterns in meaningless noise. http://www.scientificamerican.com/article/patternicity-finding-meaningful-patterns, last accessed 09 February 2016.

Sheskin, David J. (2004). Parametric and nonparametric statistical procedure. Boca Raton, FL: CRC Press.

Shilling, Chris (2004). Physical capital and situated action: A new direction for corporeal sociology. British Journal of Sociology of Education, 25(4): 473–487.

Shmueli, Galit, & Koppius, Otto R. (2011). Predictive analytics in information systems research. MIS Quarterly Executive, 35(3): 553–572.

Shneiderman, Ben (2008). Extreme visualization: Squeezing a billion records into a million pixels. Proceedings of the ACM SIGMOD International Conference on Management of Data, Vancouver, 3–12.

Siggelkow, Nicolaj (2002). Evolution toward fit. Administrative Science Quarterly, 47(1): 125–159.

Silberzahn, Raphael, & Uhlmann, Eric L. (2015). Crowdsourced research: Many hands make tight work. Nature, 526(7572): 189–191.

Silver, Nate (2012). The signal and the noise: Why most predictions fail – but some don't. New York: Penguin.

Silvermann, Rachel E. (2012). Big data upends the way workers are paid. http://www.wsj.com/articles/SB10000872396390444433504577651741900453730, last accessed 25 April 2016.

Simon, Herbert A. (1959). Theories of decision making in economics and behavioural science. American Economic Review, 49(3): 253–283.

Simon, Phil (2013). Too big to ignore – The business case for big data. Hoboken, NJ: Wiley.

Simpsons, Lorenzo C. (1995). Technology, time, and the conversations of modernity. New York: Routledge.

Sirmon, David G., Hitt, Michael A., & Ireland, R. Duane (2007). Managing firm resources in dynamic environments to create value: Looking inside the black box. Academy of Management Review, 32(1): 273–292.

Singer, Matt (2015). Welcome to the 2015 recruiter nation, formerly known as the Social Recruiting Survey. http://www.jobvite.com/blog/welcome-to-the-2015-recruiter-nation-formerly-known-as-the-social-recruiting-survey, last accessed 20 May 2016.

Smith, Adam (1776). The wealth of nations. London: Strahan and Cadell.

Smith, Merritt R., & Marx, Leo (Eds.) (1994). Does technology drive history? The dilemma of technological determinism. Cambridge, MA: MIT Press.

Smolan, Rick, & Erwitt, Jennifer (2013). The human face of big data. New York: Sterling.

Snow, John (1854). Cholera map, https://commons.wikimedia.org/wiki/File:Snow-cholera-map-1.jpg, last accessed 23 April 2016.

Society for Human Resource Management (SHRM) (2010). Background checking: conducting credit background checks, https://www.shrm.org/research/survey findings/articles/pages/creditbackgroundchecks.aspx, last accessed 27 April 2016.

Soja, Edward W. (1999). In different spaces: The cultural turn in urban and regional political economy. European Planning Studies, 7(1): 65–75.

Solove, Daniel J. (2011). Nothing to hide: The false tradeoff between privacy and security. New Haven, CA: Yale University Press.

Sood, Ashish, & Tellis, Gerard J. (2005). Technological evolution and radical innovation. Journal of Marketing, 69(3): 152–168.

Soodak, Harry, & Iberall, Arthur S. (1978). Homeokinetics: A physical science for complex systems. Science, 201(4356): 579–582.

Sorgdrager, Bas, Hulshof, Carel T., & van Dijk, Frank J. H. (2004). Evaluation of the effectiveness of pre-employment screening. International Archives of Occupational and Environmental Health, 77(4): 271–276.

Sotamaa, Olli (2010). When the game is not enough: Motivations and practices among computer game modding culture. Games and Culture, 5(3): 239–255.

Spiegelhalter, David J. (2014). The future lies in uncertainty. Science, 345(6194): 264–265.

Spier, Fred (2011). Complexity in big history. Cliodynamics, 2(1): 146–166.

Spinellis, Diomidis (2001). Fear of coding, and how to reduce it. Computer, 34(8): 98–100.

Spinney, Laura (2012). History as science. Nature, 488(7409): 24–26.

Sprague, Robert (2015). Welcome to the machine: Privacy and workplace implications of predictive analytics. Richmond Journal of Law & Technology, 21(4): 1–46.

Stacey Ralph D. (2001). Complex responsive process in organizations: Learning and knowledge creation. London: Routledge.

Stam, Cees J., van der Velden, Natascha M., Rubio, Gerard, & Verlinden, Jouke C. (2014). Redefining the role of designers within an urban community using digital design and localized manufacturing of wearables. Proceedings of the 5th International Conference on Additive Technologies (iCAT), Vienna, 82–89.

Stanley, Jay (2015). China's nightmarish citizen scores are a warning for Americans. https://www.aclu.org/blog/free-future/chinas-nightmarish-citizen-scores-are-warning-americans, last accessed 25 April 2016.

Starcraft II: Legacy of the Void (2015). Irvine, CA: Blizzard Entertainment.

Steadman, Ian (2013). Big data and the death of the theorist. http://www.wired.co.uk/news/archive/2013-01/25/big-data-end-of-theory, last accessed 09 February 2016.

Stein, Volker (2000). Emergentes Organisationswachstum: Eine systemtheoretische "Rationalisierung". München and Mering: Hampp.

Stein, Volker (2010a). Der Weg von Personalmanagement und Organisation zu Unternehmensüberlensfunktionen: Lernimpulse für pädagogische Institutionen. Arbeitspapier Nr. 001 – 2010 des Lehrstuhls für Betriebswirtschaftslehre, insb. Personalmanagement und Organisation, Universität Siegen, 10.09.2010.

Stein, Volker (2010b). Professionalisierung des Personalmanagements: Selbstverpflichtung als Weg. Zeitschrift für Management, 5(3): 201–205.

Stein, Volker (2012). Prozessorientiertes Personalmanagement – dynamische Reorientierung betrieblicher Personalarbeit. In Volker Stein, & Stefanie Müller (Eds.). Aufbruch des strategischen Personalmanagements in die Dynamisierung. Ein Gedanke für Christian Scholz (pp. 284–190). Baden-Baden and München: Nomos and Vahlen.

Stein, Volker (2013). Risk Governance – die personalwirtschaftliche Sicht. Arbeitspapier Nr. 004 – 2013 des Lehrstuhls für Betriebswirtschaftslehre, insb. Personalmanagement und Organisation, Universität Siegen, 12.06.2013.

Stein, Volker (2014). Integration in Organisationen. Revision intrasystemischer Instrumente und Entwicklung zentraler Theoreme. München and Mering: Hampp.

Stein, Volker (2015). Human resources development in times of digitalization: A dynamization agenda. Arbeitspapier Nr. 006 – 2015 des Lehrstuhls für Betriebswirtschaftslehre, insb. Personalmanagement und Organisation, Universität Siegen, 20.03.2015.

Stein, Volker, & Müller, Stefanie (Eds.) (2012). Aufbruch des strategischen Personalmanagements in die Dynamisierung. Ein Gedanke für Christian Scholz. Baden-Baden and München: Nomos and Vahlen.

Stein, Volker, & Scholz, Tobias M. (2016). Making dynamics work: The strategic potential of gamification for human resource management. 2nd Academy of Management HR Division International Conference, Sydney, 1–27.

Stein, Volker, & Wiedemann, Arnd (2016). Risk governance: Conceptualization, tasks, and research agenda. Journal of Business Economics, published online before print.

Stein, Volker, Schramm-Klein, Hanna, & Scholz, Tobias M. (2016). When ambidexterity meets informality: a hidden network versus shadow network perspective. Academy of Management Conference, Anaheim, 1–32.

Stenros, Jaakko (2015). Playfulness, play, and games. A constructionist ludogoly approach. Tampere: Tampere University Press.

Stinger, Matt (2015). Welcome to the 2015 recruiter nation, formerly known as the social recruiting survey. http://www.jobvite.com/blog/welcome-to-the-2015-recruiter-nation-formerly-known-as-the-social-recruiting-survey, last accessed 25 April 2016.

Stone, Dianna L., Deadrick, Diana L., Lukaszewski, Kimberly M., & Johnson, Richard (2015). The influence of technology on the future of human resource management. Human Resource Management Review, 25(2): 216–231.

Strong, Colin (2015). Humanizing big data: Marketing at the meeting of data, social science and consumer insight. London: Kogan Page.

Suddaby, Roy (2006). From the editors: What grounded theory is not. Academy of Management Journal, 49(4): 633–642.

Sull, Donald N., & Eisenhardt, Kathleen M. (2012). Simple rules for a complex world. Harvard Business Review, 90(9): 68–74.

Sullivan, Daniel. P., & Daniels, John D. (2008). Innovation in international business research: a call for multiple paradigms. Journal of International Business Studies, 39(6): 1081–1090.

Sullivan, Ryan, Timmermanny, Allan, & White, Halbert (1999). Data-snooping, technical trading rule performance, and the bootstrap. Journal of Finance, 54(5): 1647–1691.

Suthaharan, Shan (2014). Big data classification: Problems and challenges in network intrusion prediction with machine learning. ACM SIGMETRICS Performance Evaluation Review, 41(4): 70–73.

Sutton, Richard S. & Barto, Andrew G. (1995). Reinforcement learning: An introduction. Cambridge: MIT Press.

Sutton, Robert I., & Staw, Barry M. (1995). What theory is not. Administrative Science Quarterly, 40(3): 371–384.

Swan, Melanie (2013). The quantified self: Fundamental disruption in big data science and biological discovery. Big Data, 1(2): 85–99.

Swan, Melanie (2015). Blockchain. Blueprint for a new economy. Sebastopol, CA: O'Reilly.

Sydow, Jörg, Schreyögg, Georg, & Koch, Jochen (2009). Organizational path dependence: Opening the black box. Academy of Management Review, 34(4): 689–709.

Takebayashi, Naok, & Morrell, Peter L. (2001). Is self-fertilization an evolutionary dead end? Revisiting an old hypothesis with genetic theories and a macroevolutionary approach. American Journal of Botany, 88(7): 1143–1150.

Tallon, Paul P. (2013). Corporate governance of big data: Perspectives on value, risk, and cost. Computer, 46(6): 32–38.

Tapscott, Don, & Tapscott, Alex (2016). Blockchain revolution. How the technology behind bitcoin is changing money, business, and the world. New York: Portfolio.

Tarde, Gabriel (1893/2012). Monadology and sociology. Melbourne: re.press.

Taylor, Frederick W. (1911). The principles of scientific management. New York: Harper & Brothers.

Taylor, T. L. 2009. The assemblage of play. Games and Culture, 4(4): 331–339.

Teece, David J., Pisano, Gary, & Shuen, Amy (1997). Dynamic capabilities and strategic management. Strategic Management Journal, 18(7): 509–533.

Telgheder, Maike, & Brower-Rabinowitsch, Grischa (2016). Big-data fear plagues German healthcare. https://global.handelsblatt.com/edition/396/ressort/companies-markets/article/big-data-phobia-plagues-german-healthcare, last accessed 10 June 2016.

Tene, Omer, & Polonetsky, Jules (2012). Privacy in the age of big data: a time for big decisions. Stanford Law Review Online, 64: 63–69.

Tene, Omer, & Polonetsky, Jules (2013). Judged by the tin man: Individual rights in the age of big data. Journal on Telecommunications and High Technology Law, 11: 351–368.

Thaler, Richard, & Sunstein, Cass (2008). Nudge. Improving decisions about health, wealth, and happiness. New Haven: Yale University Press.

Thatcher, Jim (2014). Living on Fumes: Digital Footprints, Data Fumes, and the Limitations of Spatial Big Data. International Journal of Communication, 8: 1765–1783.

Thoits, Peggy. A. (1983). Multiple identities and psychological well-being: A reformulation and test of the social isolation hypothesis. American Sociological Review, 48(2): 174–187.

Thom, Norbert (2006). Entwicklungslinien – ein Beitrag zum Berufsbild des Organisators, in: Johannes-Kepler-Universität in Linz (Ed.). Dokumentation "Öffentliche und private Organisationen im Wandel (pp. 31–53). Linz: Trauner.

Thomas, Gary, & James, David (2006). Reinventing grounded theory: Some questions about theory, ground and discovery. British Educational Research Journal, 32(6): 767–795.

Thomas, Matt, & Wasmund, Sháá (2011). The smarta way to do business. Chichester: Capstone.

Thompson, James D. (1956). On building an administrative science. Administrative Science Quarterly, 1(1): 102–111.

Thorp, Jer (2012). Big data is not the new oil, https://hbr.org/2012/11/data-humans-and-the-new-oil, last accessed 26 April 2016.

Tiffin, John, & Terashima, Nobuyoshi (Eds.) (2001). HyperReality. Paradigm for the third millennium. London: Routledge.

Tilly, Charles (1984). The old new social history and the new old social history. Review (Fernand Braudel Center), 7(3): 363–406.

Todnem By, Rune (2005). Organisational change management: A critical review. Journal of Change Management, 5(4): 369–380.

Toffler, Alvin (1970). The future shock. New York: Random House.

Tokhi, Alexandros, & Rauh, Christian (2015). Die schiere Menge sagt noch nichts. Big Data in den Sozialwissenschaften. WZB Mitteilungen, 150: 6–9.

Tolman, Edward C. (1948). Cognitive maps in rats and men. Psychological Review, 55(4): 189–208.

Tourangeau, Roger, & Sternberg, Robert J. (1982). Understanding and appreciating metaphors. Cognition, 11(3): 203–244.

Townley, Barbara (1993). Foucault, power/knowledge, and its relevance for human resource management. Academy of Management Review, 18(3): 518–545.

Training Dummies (2016). Episode 122 – Meet the sims. http://www.thetrainingdummies.com/2016/03/04/episode-122-meet-the-sims, last accessed 11 May 2016.

Transfermarkt (n.d.) Aufstellung FC Liverpool – AC Mailand, http://www.transfermarkt.de/fc-liverpool_ac-mailand/aufstellung/spielbericht/68267, last accessed 25 April 2016.

Traub, Amy (2013) Credit reports and employment: findings from the 2012 national survey on credit card debt of low-and middle-income households. Suffolk University Law Review, 46: 983–995.

Troester, Mark (2012). Big data meets big data analytics. Three key technologies for extracting real-time business value from the big data that threatens to overwhelm traditional computing architectures. SAS White Paper: 1–10.

Trzebski, Andrzej (1994). Homeodynamics versus homeostasis: periodicities superimposed on non-linear dynamic sympathetic tone generated in ventral medulla. Acta Neurobiologiae Experimentalis, 54(2): 109–125.

Tucker, Patrick (2013). Has big data made anonymity impossible? MIT Technology Review Business Report, 64–66.

Tumasjan, Andranik, Sprenger, Timm O., Sandner, Philipp G., & Welpe, Isabell M. (2010) Predicting elections with twitter: What 140 characters reveal about political sentiment. Proceedings of the Fourth International AAAI Conference on Weblogs and Social Media, Washington D.C., 178–185.

Turchin, Peter (2008). Arise 'cliodynamics'. Nature, 454(7200): 34–35.

Turchin, Peter (2012). Psychohistory and cliodynamics. https://evolution-institute.org/blog/psychohistory-and-cliodynamics, last accessed 25 April 2016.

Turing, Alan M. (1950). Computing machinery and intelligence. Mind, 59(236): 433–460.

Tversky, Amos, & Kahneman, Daniel (1973). Availability: A heuristic for judging frequency and probability. Cognitive Psychology, 5(2): 207–232.

Tversky, Amos, & Kahneman, Daniel (1974). Judgment under uncertainty: Heuristics and biases. Science, 185(4157): 1124–1131.

U.S. Congress (1961). Developments in the field of detection and identification of nuclear explosions (Project Vela) and relationship to test ban negotiations: Hearings before the joint committee on atomic energy. Washington, D.C.: U.S. Government Printing Office.

U.S. Senate (1972). Agriculture-environmental and consumer protection appropriations for fiscal year 1972: Hearings before a subcommittee of the committee on appropriations. Washington, D.C.: U.S. Government Printing Office.

Ulrich, Dave, Younger, Jon, Brockbank, Wayne, & Ulrich, Michael D. (2013). The state of the HR profession. Human Resource Management, 52(3): 457–471.

Umpleby, Stuart A. (1990). The science of cybernetics and the cybernetics of science. Cybernetics and Systems: An International Journal, 21(1): 109–121.

Umpleby, Stuart A., & Dent, Eric B. (1999). The origins and purposes of several traditions in systems theory and cybernetics. Cybernetics & Systems, 30(2): 79–103.

Uprichard, Emma (2013). Focus: Big data, little questions. Discover Society, (1): 1–6.

Uribe-Jongbloed, Enrique, & Espinosa-Medina, Hernán D. (2014). A clearer picture: Towards a new framework for the study of cultural transduction in audiovisual market trades. Observatorio (OBS*), 8(1): 23–48.

Valiant, Leslie G. (1984). A theory of the learnable. Communications of the ACM, 27(11): 1134–1142.

Valiant, Leslie G. (2013). Probably approximately correct. New York: Basic Books.

van Asselt, Marjolein B. A., Renn, Ortwin (2011) Risk governance. Journal of Risk Research 14(4): 431–449.

van Dijck, José (2014). Datafication, dataism and dataveillance: Big data between scientific paradigm and ideology. Surveillance & Society, 12(2):197–208.

van Dijck, José, & Poell, Thomas (2013). Understanding social media logic. Media and Communication, 1(1): 2–14.

van Dijk, Teun A. (1998). Opinions and ideologies in the press. In Allen Bell, & Peter Garrett (Eds.). Approaches to media discourse (pp. 21–63). Oxford: Blackwell.

van Inwagen, Peter (1983). An essay on free will. Oxford: Oxford University Press.

van Rijmenam, Mark (2013). Why the 3V's are not sufficient to describe big data. https://datafloq.com/read/3vs-sufficient-describe-big-data/166, last accessed 03 February 2016.

van Valen, Leigh (1973). A new evolutionary law. Evolutionary Theory, 1: 1–30.

Varian, Hal R. (2014). Big data: New tricks for econometrics. Journal of Economic Perspectives, 28(2): 3–28.

Vaux, David L. (2013). Statistics: Number-crunching in the raw. Nature, 493(7432): 301.

Veblen, Thorstein (1921/2001). The engineers and the price systems. Kitchener, Ontario: Batoche Books.

Vidgen, Bertie, & Yasseri, Taha (2016). P-values: misunderstood and misused. arXiv preprint arXiv:1601.06805.

Vigen, Tyler (2015). Spurious correlations. Correlation does not equal causation. New York: Hachette Books.

Visser, Wayne (2011). The age of responsibility: CSR 2.0 and the new DNA of business. Chisester, GB: Wiley.

von Bertalanffy, Ludwig (1965). Zur Geschichte theoretischer Modelle in der Biologie. Studium Generale, 8: 290–298.

von Bertalanffy, Ludwig (1968). General system theory. New York: Braziller.

von Bertalanffy, Ludwig (1972). The history and status of general systems theory. Academy of Management Journal, 15(4): 407–426.

von Clausewitz, Carl (1832/1976). On war. Princeton: Princeton University Press.

von Foerster, Heinz (1979). Cybernetics of cybernetics. In Klaus Krippendorf (Ed.). Communication and control (pp. 5–8). New York: Gordon and Breach.

von Foerster, Heinz (2003). Understanding understanding. Essays on cybernetics and cognition. New York: Springer.

von Glasersfeld, Ernst (1979). Cybernetics, experience, and the concept of self. In Mark N. Ozer (Ed.). A cybernetic approach to the assessment of children: Toward a more humane use of human beings (pp. 67–113). Boulder, CO: Westview.

von Glasersfeld, Ernst (1995). Radical Constructivism. London: Falmer.

Voss, Peter (2007). Essentials of general intelligence: The direct path to artificial general intelligence. In Ben Goertzel, & Cassio Pennachin (Eds.). Artificial general intelligence (pp. 131–157). Berlin: Springer.

Wadhwa, Anu, & Kotha, Suresh (2006). Knowledge creation through external venturing: Evidence from the telecommunications equipment manufacturing industry. Academy of Management Journal, 49(4): 819–835.

Waldrop, Mitchell M. (1993). Complexity: The emerging science at the edge of order and chaos. New York: Simon and Schuster.

Walker, Joseph (2012a). Moneyball and the HR department. http://blogs.wsj.com/digits/2012/04/16/moneyball-and-the-hr-department, last accessed 26 April 2016.

Walker, Joseph (2012b). Meet the new boss: Big data. http://www.wsj.com/articles/SB10000872396390443890304578006252019616768, last accessed on 16 September 2013.

Walter, Chip (2005). Kryder's law. Scientific American, 293(2): 32–33.

Wang, Catherine L., & Ahmed, Pervaiz K. (2007). Dynamic capabilities. A review and research agenda. International Journal of Management Reviews, 9(1): 31–51.

Wang, Rui, Chen, Fanglin, Chen, Zhenyu, Li, Tianxing, Harari, Gabriella, Tignor, Stefanie, Zhou, Xia, Ben-Zeev, Dror, & Campbell, Andrew T. (2014). StudentLife: Assessing mental health, academic performance and behavioral trends of college students using smartphones. Proceedings of the ACM International Joint Conference on Pervasive and Ubiquitous Computing, Seattle, 3–14.

Ward, Jonathan S., & Barker, Adam (2013). Undefined by data: A survey of big data definitions. arXiv preprint arXiv:1309.5821.

Wason, Peter C. (1960). On the failure to eliminate hypotheses in a conceptual task. Quarterly Journal of Experimental Psychology, 12(3): 129–140.

Watch Dogs (2014). Montreal: Ubisoft.

Watson, David (1982). The actor and the observer: How are their perceptions of causality divergent? Psychological Bulletin, 92(3): 682–700.

Webb, Stephen (2002). If the universe is teeming with aliens ... where is everybody? Fifty solutions to the Fermi paradox and the problem of extraterrestrial life. New York: Copernicus Books.

Weber, Max (1919). Wissenschaft als Beruf. Erweiterte Fassung des Vortrags beim Freistudentischen Bund Landesverband Bayern.

Weber, Max (1952). The Protestant ethic and the spirit of capitalism. New York: Scribner.

Weick, Karl E. (1976). Educational organizations as loosely coupled systems. Administrative Science Quarterly, 21(2): 1–19.

Weick, Karl E. (1979). The social psychology of organizing. New York: McGraw-Hill.

Weick, Karl E. (1982). Administering education in loosely coupled schools. The Phi Delta Kappan, 63(10): 673–676.

Weick, Karl E. (1989). Theory construction as disciplined imagination. Academy of Management Review, 14(4): 516–531.

Weick, Karl E. (1993). Organizational redesign as improvisation. In George. P. Huber, & William H. Glick (Eds.). Organizational change and redesign (pp. 346–379). Oxford: Oxford University Press.

Weick, Karl E. (1995). What theory is not, theorizing is. Administrative Science Quarterly, 40(3): 385–390.

Weinberger, David (2011). Too big to know. New York: Basic Books.

Weinberger, David (2013). Die digitale Glaskugel. In Heinrich Geiselberger, & Tobias Moorstedt (Eds.). Big Data. Das neue Versprechen der Allwissenheit (pp. 219–237). Berlin: Suhrkamp.

Weiss, Sholom, M., & Indurkhya, Nitin (1998). Predictive data mining: A practical guide. San Francisco: Morgan Kaufmann.

West, Geoffrey (2013). Big data needs a big theory to go with it. http://www.scientificamerican.com/article/big-data-needs-big-theory, last accessed 25 April 2016.

West, Jonathan P., & Bowman, James S. (2014). Electronic surveillance at work. An ethical analysis. Administration & Society, published online before print.

Weyrich, Michael, Schmidt, Jan-Philipp, & Ebert, Christof (2014). Machine-to-machine communication. IEEE Software, 31(4): 19–23.

Whetten, David A. (1989). What constitutes a theoretical contribution? Academy of Management Review, 14(4): 490–495.

Whitehead, Alfred N. (1929). Process and reality. An essay in cosmology. Gifford lectures delivered in the University of Edinburgh during the session 1927–1928. New York: Macmillan.

Whitfield, John (2005). Complex systems: order out of chaos. Nature, 436(7053): 905–907.

Whitson, Jennifer R. (2013). Gaming the quantified self. Surveillance & Society, 11(1/2): 163–176.

Wiener, Norbert (1948). Cybernetics or control and communication in the animal and the machine. New York: Wiley.

Wild, Chris J. (1994). Embracing the "wider view" of statistics. The American Statistician, 48(2): 163–171.

Wild, Chris J., & Pfannkuch, Maxine (1999). Statistical thinking in empirical enquiry. International Statistical Review, 67(3): 223–248.

Wilden, Ralf, Gudergan, Siegfried, & Lings, Ian (2010). Employer branding: Strategic implications for staff recruitment. Journal of Marketing Management, 26(1–2): 56–73.

Williams, Wendell (2013). The big data HR fad. http://www.ere.net/2013/06/19/the-big-data-hr-fad, last accessed 27 April 2016.

Wilson, Marie, Chen, Shaohui, & Erakovic, Ljiljana (2006). Dynamics of decision power in the localization process: comparative case studies of China-Western IJVs. International Journal of Human Resource Management, 17(9): 1547–1571.

Wilson, Robert C. (2015). 10 science fiction writers predict how our world will change in the next 10 years. http://www.huffingtonpost.com/2015/05/06/sci-fi-predictions_n_ 7102742.html, last accessed 06 January 2016.

Wing, Jeannette M. (2006). Computational thinking. Communications of the ACM, 49(3): 33–35.

Winner, Langdon (1993). Upon opening the black box and finding it empty: Social constructivism and the philosophy of technology. Science, Technology, & Human Values, 18(3): 362–378.

Winner, Langdon (2003). Social constructivism: Opening the black box and finding it empty. In Robert C. Scharff, & Val Dusek (Eds.). Philosophy of technology: The technological condition: An anthology (pp. 233–244). Malden, MA: Blackwell Publishers.

Winner, Langdon (2004). Technology as forms of life. In David M. Kaplan (Ed.). Readings in the philosophy of technology (pp. 103–113). Oxford: Rowman & Littlefield.

Wittgenstein, Ludwig (1922). Tractatus logico-philosophicus. London: Kegan Paul, Trench, Trubner.

Wolf, Gary (2010). The data-driven life. http://www.nytimes.com/2010/05/02/magazine/02self-measurement-t.html?_r=0, last accessed 25 April 2016.

World of Warcraft (2004). Irvine, CA: Blizzard Entertainment.

Wright, Alex (2014). Big data meets big science. Communications of the ACM, 57(7): 13–15.

Wright, Jason T., Mullan, Brendan, Sigurdsson, Steinn, & Povich, Matthew S. (2014). The Ĝ infrared search for extraterrestrial civilizations with large energy supplies. I. Background and justification. Astrophysical Journal, 792(1): 1–16.

Wright, Jason T., Cartier, Kimberly M. S., Zhao, Ming, Jontof-Hutter, Daniel, & Ford, Eric B. (2016). The Ĝ search for extraterrestrial civilizations with large energy supplies. IV. The signatures and information content of transiting megastructures. Astrophysical Journal, 816(1): 1–22.

Wright, Sewall (1932). The roles of mutation, inbreeding, crossbreeding, and selection in evolution. Proceedings of the Sixth International Congress on Genetics, 355–366.

Wyss-Flamm, Esther D., & Zandee, Danielle P. (2001). Navigating between finite and infinite games in the managerial classroom. Journal of Management Education, 25(3): 292–307.

Yates, F. Eugene. (1994). Order and complexity in dynamical systems: homeodynamics as a generalized mechanics for biology. Mathematical and Computer Modelling, 19(6): 49–74.

Yeung, Karen (2016). 'Hypernudge': Big data as a mode of regulation by design. Information, Communication & Society, published online before print.

Yudkowsky, Eliezer (2008). Cognitive biases potentially affecting judgment of global risks. In Nick Bostrom, & Milan M. Ćirković (Eds.). Global catastrophic risks (pp. 91–119). Oxford: Oxford University Press.

Yuhas, Alan (2014). Rosetta: What went right with Philae, what went wrong and how it can be fixed. https://www.theguardian.com/science/2014/nov/13/rosetta-philae-comet-mission-what-went-right-what-went-wrong-and-how-it-can-be-fixed, last accessed 25 April 2016.

Zanoni, Patrizia, & Janssens, Maddy (2004). Deconstructing difference: The rhetoric of human resource managers' diversity discourses. Organization Studies, 25(1): 55–74.

Zarsky, Tal (2008). Law and online social networks: Mapping the challenges and promises of user-generated information flows. Fordham Intellectual Property, Media and Entertainment Law Journal, 18(3): 741–783.

Zhang, Yanxia, & Zhao, Yongheng (2015). Astronomy in the big data era. Data Science Journal, 14: 1–9.

Zichermann, Gabe, & Cunningham, Christopher (2011). Gamification by design: Implementing game mechanics in web and mobile apps. Sebastopol, CA: O'Reilly.

Zimmerman, Brenda J., & Hurst, David K. (1993). Breaking the boundaries. Journal of Management Inquiry, 2(4): 334–355.

Zins, Chaim (2007). Conceptions of information science. Journal of the American Society for Information Science and Technology, 58(3): 335–350.

Zittrain, Jonathan L. (2006). The generative internet. Harvard Law Review, 119(7): 1974–2040.

Zuboff, Shoshana (1988). In the age of the smart machine. The future of work and power. New York: Basic Books.

Zuboff, Shoshana (2014). The human factor. http://www.faz.net/aktuell/feuilleton/debatten/the-digital-debate/digital-economy-the-human-factor-13050472.html, last accessed 09 February 2016.

Zwitter, Andrej (2014). Big data ethics. Big Data & Society, 1(2): 1–6.

Zyskind, Guy, Nathan, Oz, & Pentland, A. Sandy. (2015). Decentralizing privacy: Using Blockchain to protect personal data. IEEE Security and Privacy Workshops (SPW), 180–184.